W9-CBF-542

Also available in the Bloomsbury Sigma series:

THE VINYL FRONTIER

THE STORY OF THE
VOYAGER GOLDEN RECORD

Jonathan Scott

BLOOMSBURY SIGMA
LONDON · OXFORD · NEW YORK · NEW DELHI · SYDNEY

BLOOMSBURY SIGMA
Bloomsbury Publishing Plc
50 Bedford Square, London, WC1B 3DP, UK

BLOOMSBURY, BLOOMSBURY SIGMA and the Bloomsbury Sigma logo
are trademarks of

Bloomsbury Publishing Plc

First published in the United Kingdom in 2019

Copyright © Jonathan Scott, 2019

Jonathan Scott has asserted his right under the Copyright, Designs and Patents
Act, 1988, to be identified as Author of this work

All rights reserved. No part of this publication may be reproduced or transmitted
in any form or by any means, electronic or mechanical, including photocopying,
recording, or any information storage or retrieval system, without prior
permission in writing from the publishers

Bloomsbury Publishing Plc does not have any control over, or responsibility for,
any third-party websites referred to or in this book. All internet addresses given
in this book were correct at the time of going to press. The author and publisher
regret any inconvenience caused if addresses have changed or sites have ceased to
exist, but can accept no responsibility for any such changes

A catalogue record for this book is available from the British Library

Library of Congress Cataloguing-in-Publication data has been applied for

ISBN: HB: 978-1-4729-5613-2
eBook: 978-1-4729-5611-8

2 4 6 8 10 9 7 5 3

Typeset in Deanta Global Publishing Services, Chennai, India
Printed and bound in Great Britain by CPI Group (UK) Ltd, Croydon CR0 4YY

Bloomsbury Sigma, Book Forty-three

MIX
Paper from
responsible sources
FSC
www.fsc.org   FSC® C020471

To find out more about our authors and books visit www.bloomsbury.com and
sign up for our newsletters

For the *other* book club

Contents

Prologue

I still remember the first mixtape someone made for me. It had three words written on the side in blue ink: 'Punk for Jono'.

This pre-playlist playlist was produced in around 1987, 10 years after the launch of the Voyager space probes. Unlike the Voyager interstellar message, it wasn't imprinted on a golden record encased in an aluminium cover and attached to the side of a spacecraft. It was a plastic C90 tape handed to me by Ed, a cool kid in the year above. The only thing it had in common with the Voyager records was that it included music selected by human hand. I would have been 12. I had stopped listening to *Starlight Express*, noting that Lloyd Webber musicals didn't carry much weight in the school cool-o-sphere, and had recently discovered The Blues Brothers, *Atlantic Soul Classics* and Otis Redding.

Then Ed made me a tape.

He only recorded one side of the tape, and there was quite a long gap at the end, so it can only have been around 40 minutes in length. The tape has not survived, but my memory of it remains clear.

Ed set the record levels a little too high. So by the time Topper Headon's rumbling intro to 'I Fought the Law' had crescendoed, the drums were distorting. Then came the power chords. It was indescribably thrilling. Let's just say Arthur Conley's 'Sweet Soul Music' suddenly seemed a little thin.*

Ed's compilation taught me the importance of starting strong. And even though I haven't actually seen this particular tape for 15 years, I can still recall every song – two more by The Clash ('Safe European Home', 'Guns of Brixton'), 'Guns for the Afghan Rebels' by Angelic Upstarts, 'Greatest Cockney

---

* Although I would like to make it clear that I still really love 'Sweet Soul Music'.

Rip Off' by Cockney Rejects, 'Seattle' by PiL, 'Eton Rifles' by The Jam, 'Anarchy in the UK' by the Sex Pistols and The Damned covering 'Jet Boy, Jet Girl'. This was the first mixtape someone made for me, and in a way it remains unsurpassed.

Around the same time I received a double tape deck for my birthday. Suddenly, at my disposal, was the heady power of being able to share and disseminate music. I was able to easily transfer any music I had from one tape to another. I'm pretty confident I owe Whitesnake quite a lot of money. I must have supplied copies of 'Here I Go Again' (from *Now That's What I Call Music 10*) to most of the boys in my dormitory. But the first proper compilation, mixtape, playlist – whatever you want to call it – that I made for someone else, came much later. It was for a girl called Beth.

It was the summer term of my penultimate year at boarding school. I had just emerged from a four-year imprisonment in braces. I had joined a band. I had got drunk. I had successfully snogged one female to date. Then a friend informed me that Beth had taken a shine to me. That was the summer I found out a little of what love was like. I'd had crushes before – Fairuza Balk in *Return to Oz*, for example – but nothing like this. I didn't know her well. She was a couple of years younger than me. But she was pretty and funny, with a disarming, slightly dozy way of moving and talking, as if her body was suspended from a head full of helium.

Messages were exchanged. We were to meet at Doctor's Lake after lunch. We had the whole afternoon. Doctor's Lake was in reality a fairly dismal pond, but it was a place within school grounds where couples might go with a relatively high expectation of solitude. We went there. We were alone. I remember it being hot. I remember being terrified of any physical contact. I remember leaning back, accidentally touching her hand, and actually screaming with the shock.

The afternoon wore on. Screaming aside, there had been no physical contact. It was becoming awkward. We had to go back to our respective boarding houses soon. Eventually we did, walking in a slow meandering way, conscious of important work still to attend to, expectations to be met.

Then, as we approached the back of the old squash courts with the leaky roof, we faced each other. I pushed through a bewildering fog of terror, and we managed a kiss.

That evening, snog under my belt, I set about fulfilling a promise I'd made earlier in the day – I would make her a tape. I can still recall the feeling, the excitement, the compulsion to create the ultimate calling card of cool. In that C90 tape, I was seeking to represent myself as a sensitive, discerning boy with startlingly good taste.

There was a time when I kept a journal of every tape I made, with each song noted down so as to guard against that cardinal sin of giving someone the same song twice. Although that journal no longer survives, I still remember some highlights from the first tape I made Beth. This was around 1991 and I was a British indie kid, so there was certainly 'Vapour Trail' by Ride, 'Grey Cell Green' by Ned's Atomic Dustbin, some Violent Femmes (probably 'Good Feeling'), and I remember very clearly that sides one and two were kicked off by songs from earnest Irish rockers Power of Dreams.

From that day forwards, for the best part of two decades, any good friend of either sex would probably receive at least one compilation from me. Some were short, on C60 tapes, some were long. At times I ventured into double and triple albums (reusing those tape cases that generally housed cheap blues compilations). I gave the tapes names and made covers, usually with random punk writing from letters cut out of magazines. Later, I made the leap to producing compilations on CD, and even playlists that only existed in digital form. But it was never quite the same.

Back in my mixtape heyday I had rules: I stopped including more than one song by the same artist or band, and all the songs had to be unknown to the receiver (although this wasn't always possible). I also liked contrasts. I liked to follow serious, mournful songs with novelty pop. I liked to have quiet followed by loud, or loud by quiet, so a song from Anthrax, for example, might well be followed by Flanders and Swann.

Another important rule was that I had to minimise space at the end of each side of the cassette. I gathered together a host of short songs ('Wienerschnitzel' by Descendents, for instance) that came in useful when trying to fill each side of the tape right to its audio brim. If the dreaded snap of the 'record' button popping up occurred even a second before the final chords of the song faded away, or conversely if there was too long a pause afterwards, it would not do. It had to fit, perfectly, with only seconds to spare.

Why am I telling you all this? Because I am not a scientist. I've always loved astronomy, sure. As a kid I watched and re-watched documentaries on Halley's Comet, on Voyager 2, on Neptune. I kept a scrapbook of articles culled from *Scientific American* and *New Scientist*. I know the name of Jupiter's fourth-largest moon. I understand the justification for Pluto's declassification. I know what a pulsar is. But, despite a fascination with astronomy, I don't have more than a layman's grasp of what's going on out there. I do, though, have an understanding of what it's like to make a compilation.

When I handed Beth that tape in the early summer of 1991, I was trying to share my soul, to create an image of me, encapsulated in 90 minutes of music. When a group of scientists, artists and writers gathered in Ithaca, New York, to begin work on the Voyager Golden Record, they were attempting to capture the soul of humanity in 90 minutes of music.

Try to imagine it. Think of the pressure. What would you choose? You have an hour and a half to represent Earth's music. Plus, if we assume that you're going to make the very sensible decision to include the first-pressing Stone Roses LP in its entirety, you really only have around 30 minutes left to play with.

This is a story of the summer of 1977 – when science rubbed up against art to create a monument that will, in all probability, outlive us all. When we are dust, when the Sun dies, these two golden analogue discs, with their handy accompanying stylus and instructions, will still be speeding off further into the cosmos. And alongside their music, photographs and data, the discs will still have etched into their fabric the sound of one woman's brainwaves – a recording made on 3 June 1977, just weeks before launch. The sound of a human being in love with another human being. Just as I was in love with Beth.

CHAPTER ONE

The Naked Pioneers

'Ad astra per aspera' – *'Through difficulties to the stars'*
Carl Sagan

Two space probes called Voyager 1 and Voyager 2 were sent to the outer planets in the late 1970s to beam back lots of lovely images and data of the gas giants and their moons. Primary mission complete, and with no way of being controlled, the probes were doomed to drift forever in the unimaginable void of interstellar space.

With this gloomy-sounding outcome in mind, NASA decided to do something very optimistic. They would send with them a message, on the very slim chance that they would one day be recovered by some little green chap. The message took the form of a metal record. A record designed to convey something about our origins, our civilisation, our art, through sounds, images and music. *CliffsNotes to Earthlings.*

To tell the story of this record, its creators and the people who chose that moment to fall in love, we need to begin by backing up a little. You see, *Voyager Golden Record* is a sequel. It's the more ambitious, bigger-budget sequel to *Pioneer Plaque.* Five years before Voyager, the Pioneer probes became the first man-made objects destined to reach interstellar space, to cast off the gravitational cloak of our solar system and head out into it forever. While this was known to everyone involved in the mission, it took an outsider looking in – a writer for *Christian Science Monitor* – to draw attention to the magnitude of this fact. Pioneer 10 and 11 were to be humankind's first emissaries to the stars. Here was a first-time opportunity to send a message, a greeting to any intelligent beings who might chance upon them.

Armed with enthusiasm and a deadline of just three weeks, a three-strong team, comprising an artist, an astronomer and

an astrophysicist, thrashed out a design for a modest metal
plaque, complete with star map and naked human figures.
However, one of the strangest aspects of the Pioneer
plaques – the first of which was hurled towards Jupiter in
March 1972 – was the absence of vulva.

The last-minute sanitising of the female figure, for fear of
offending a domestic audience, captures an essence of the
environment in which the Pioneer plaques were forged.
These are not only messages to some imagined future alien
interaction, they are also time capsules, snapshots of when
they left Earth. The removal of the vulva is a lens through
which we can view America nearly 50 years ago. Despite the
swinging sixties, despite mass movements, protests, social
reforms and upheavals, despite convention-busting cinema
coming out of the American New Wave, America in 1972
remained a conservative environment. NASA was in the
glare of the public gaze, at the mercy of popular opinion. It
was funded by American tax dollars, and everyone knew it.
And to tell the story of the missing vulva, we need to back up
just a little bit further.

In 1964 a graduate student working at NASA's Jet
Propulsion Laboratory* noticed something important. His
name was Gary Flandro. He had finished school in 1957,
just before Sputnik, perfectly timing his arrival at JPL with
the start of the space race. The first of the Pioneer missions
took place the following year, when a probe designed to
achieve moon orbit, failed some 73.6 seconds after launch.
A steady stream continued to punch holes in the atmosphere.
Some were lost at launch, others failed to reach their

---

* The Jet Propulsion Laboratory (JPL) began life in the mid-1930s as
an isolated haven at the foot of the San Gabriel Mountains where
California Institute of Technology (Caltech) students could blow
stuff up without hurting anybody. During the Second World War it
was commandeered by the Army. Then, after Sputnik signalled
Russia's round-one victory in the space race, it was where America
set about designing and building their counterpunch: Explorer 1,
America's first satellite.

desired orbit, others did very well. A personal favourite is Pioneer 5 from March 1960 – a spherical probe sent to measure magnetic forces in the space between the orbits of Earth and Venus – that looks very similar to the intimidating black ball that approaches Princess Leia in *Star Wars: A New Hope*.

Flandro had been investigating the knotty problem of how an object might be sent further, towards the outer planets. The general consensus at the time was that this was virtually impossible without something called gravity assist. Jupiter held the keys to the outer planets. Without Jupiter's gravity, any object sent in that direction would eventually fall back to Earth's orbit. But with Jupiter there, an object could fly past, pick up an enormous boost of energy, and then be hurled out towards Saturn. Then, in theory, it could do the same thing at Saturn, and head off further towards Uranus, Neptune and so on.

While working on trajectories, Flandro realised something mind-blowing: that by the late 1970s all the outer planets would be on the same side of the Sun. This alignment of Jupiter, Saturn, Uranus and Neptune could enable a single craft to visit all four outer planets by using gravity assists in what would be dubbed the 'Grand Tour'. This wasn't quite a single one-time option – there were various possible iterations and trajectories – but such an advantageous planetary arrangement would not occur again for another 175 years. Flandro had uncovered a chance to explore a number of planets in one go, at a fraction of the cost. Suddenly NASA had been given the mother of all deadlines.

★★★

To plan for a planetary Grand Tour, NASA needed Pioneers 10 and 11 to dip their toes in the waters of the outer solar system, to see if such an endeavour was even possible. The idea to include a message with the Pioneers came relatively late in the day. Eric Burgess, an English freelancer, had been writing about space missions since 1957. He's the one who

held up his hand in class and pointed out to the world that
NASA was about to throw something further than anything
had ever been thrown. And during conversations with writers
Richard Hoagland and Don Bane the idea for attaching some
kind of physical message to the craft was forged. Burgess was
unsure about pitching direct to NASA, so instead he
approached Carl Sagan, whose eyes lit up.

Sagan was already a well-known astronomer, with a
growing public profile, but he was not yet the household
name he would become. He was a coal-face scientist, with
acknowledged achievements, enthusiasms and specialisations
under his belt. He would study the greenhouse effect on
Venus, seasonal dust patterns on Mars, the environment of
Saturn's moon Titan. He would play a medal-winning role in
NASA's Mariner 9 mission, and contribute to the Viking,
Voyager and Galileo missions.

In early December 1971, astronomers and astrophysicists
descended on San Juan, Puerto Rico, for the 136th meeting
of the American Astronomical Society. During a coffee break
Sagan began talking to his friend and colleague Frank Drake
about the idea for a message aimed at extraterrestrials. He was
preaching to the converted – Frank had been thinking about
communicating with aliens for years.*

---

* In 1959 a paper had appeared in *Nature* under the heading
'Searching for Interstellar Communications'. This was the cosmic
equivalent of pioneering jazz guitarist Charlie Christian's 'Guitarmen,
Wake Up and Pluck!' article, printed some 20 years before. Much as
Christian had inspired a generation of players to electrify and amplify
their instruments, so the *Nature* paper put forward a realistic strategy
for searching for extraterrestrial intelligence. It pointed out that
scientists now had equipment within their grasp that could scan for
radio signals from any alien civilisation with comparable technology.
Written by Giuseppe Cocconi and Philip Morrison, the paper ends:
'The probability of success is difficult to estimate; but if we never
search, the chance of success is zero.' And Frank Drake became the
first radio astronomer to do exactly that, with an experiment named
'Project Ozma' in 1960.

Writing in 'The Foundations of the Voyager Record' in 1978, Drake recalled that many of the parameters for the Pioneer plaques were pinned down during that frantic coffee break in Puerto Rico. Before the end of recess, Sagan and Drake had decided they wanted to convey where, when and who the craft came from, without using written language.

To begin with, Sagan suggested a map showing binary stars and constellations, but Drake had a more elegant solution. He suggested a map showing Earth in relation to certain pulsars – spinning neutron stars that emit regular beams of radiation. He reasoned that, as the frequencies of the pulses shift slowly over time, it should be possible for some future being to calculate – from the difference in the frequency of those pulses as recorded on the plaque, compared to the pulses at that future time – the approximate date when the ship's plaque was designed.

Sagan had the connections and influence within NASA, but it wasn't an easy sell. Even the smallest addition of weight to Pioneer 10 would throw off NASA's calculations. To enable Sagan's project to proceed, NASA would have to unclench, let go the reins, and okay its engineers to re-compute the figures. After weeks of delays and bureaucratic wrangling, Sagan had his green light. But with the launch fast approaching, it was a green light with a strict deadline of just three weeks.

Drake set to work on his pulsar map, while Sagan drew up a second diagram showing the solar system and Pioneer's trajectory. They had covered the when and the where. Now they needed to tackle the who. Step forwards the third collaborator: artist Linda Salzman Sagan, who had been married to Carl since April 1968.

Salzman began working on simple line drawings of human figures, expressly designed to carry physical characteristics that crossed racial boundaries. The team ruled out giving the humans clothes, discounted diagrams showing veins, muscles, lungs or organs, and abandoned an early idea of showing them holding hands – the viewer might interpret the image as being of a single organism.* The final design shows a man and woman drawn to scale before a schematic of the Pioneer spacecraft. The man is slightly taller than the woman, his arm raised in greeting. And they're both naked.

Robert Kraemer, Director of Planetary Programmes at NASA, was at the meeting called in January 1972 to approve the designs. He recalled feeling nervous about the images. He could see Linda had done a good job, yet he feared a public reaction to the nudity. His boss, John Naugle, shared some but not all of his anxiety, okaying the design with the proviso that they erase the short line that indicated the vulva.

The plaque was rubber-stamped. The design was engraved on a six-by-nine-inch gold anodised aluminium plate and bolted to Pioneer 10 in a place that would protect it from the worst of the interstellar dust. It was ready to go.

The existence of the plaque was revealed to the public in February 1972. Both Sagan and Drake fielded press interviews, posed for pictures and talked before cameras[†] next to the soon-to-be-launched vessel. There was plenty of enthusiasm and excitement, but there were several strands of negativity and derision, most of it aimed squarely at Salzman's human figures. Some objected to a skewed gender

---

* A NASA post-mission report, penned by our three collaborators, took a cautious tone: 'It is not clear how much evolutionary or anthropological information can be deduced from such a sketch drawing … It seems likely, if the interceptor society has not had previous contact with organisms similar to human beings, that many of the body characteristics shown will prove deeply mysterious.'

† Frank recalls: 'I had to go on a talk show in Toronto. I was on for maybe five minutes or something. I thought, "Oh, this went very well." I thought I was articulate, I didn't mumble. And all I got was this horrified look from people. They were in shock, they were speechless. "What is it? What happened? What's wrong?" I asked. And they said it was the first time a naked human had ever been shown on Canadian television. "We're all going to be fired!" Luckily nothing happened.'

bias, others to the figures' racial make-up. That only three people worked on the message also caused disquiet in some quarters. But the vast majority were simply scandalised by the nakedness. What was NASA doing, spending our taxes to send this filth out into the universe? Cartoonists had a field day, lampooning NASA as interstellar pornographers. Even those covering the story had to tread carefully. The *Chicago Sun-Times* editors avoided trouble by airbrushing out the 'sexual' parts of the image, removing more and more through subsequent editions of the day. The *Philadelphia Inquirer* upheld standards by covering the woman's nipples and man's genitals. When the *Los Angeles Times* showed the design in all its naked glory, it attracted letters of complaint. There was even one voice who felt that the male figure was performing a Nazi salute.

Pioneer 10 left this clamour far behind, launching from Cape Canaveral, Florida on 2 March 1972. The mission was to be a barnstorming success. The bare facts are these: between July 1972 and February 1973 it became the first spacecraft to traverse the asteroid belt – itself seen as a significant achievement. The first of around 500 photographs of Jupiter was beamed back in November 1973. The closest approach (132,252km) took place in December 1973, by which time it was being followed by its equally successful sister, Pioneer 11, which launched in April 1973 and became the first craft to visit Saturn. Primary missions complete, both headed off into space, their instruments taking the environmental pulse of the far reaches of the solar system and the heliosphere.

Radio communications with Pioneer 10 were finally lost in 2003 – electric power was by then too weak for Earth's receivers to pick up any transmission. At that point the probe was some 12 billion km, or around 80 Astronomical Units, from Earth. For many years it held the gold medal for the most remote Earth-made object, until it was surpassed in February 1998 by the faster Voyager 1. And as I write, Pioneer 10 is heading towards the constellation Taurus at a speed (relative to the Sun) of about 12km a second.

Back on Earth, the plaque's three-strong creative team were left to ruminate over their bruising encounter. There had been bemused surprise and some alarm at the public reaction, but they came to view the majority of the criticisms as unfounded, and were also pleased that it had provoked such interest. Drake wrote in 'Foundations' that the experience had taught them humility in their approach to the future Voyager enterprise.

Today you can quickly find digital threads and articles discussing the absent vulva, still picking over the size and stance of the female figure in relation to the male. It's certainly possible to see how the image can be interpreted as being of a docile woman standing beside the active man. Look at the way her head is turned slightly towards him, while he manfully stares directly at the viewer. He seems to be saying: 'Stand back, darling, let me handle the important business of waving.' It's all rather 1972. When *The Spokesman-Review* newspaper ran with the story, it appeared next to a one-paragraph news item titled 'Women Get In', reporting that Oxford University's Sporting Club was about to admit women for the first time in 109 years ('albeit for a trial period on Saturday nights only') and directly above a quarter-page advertisement for the Preliminary Miss USA Pageant.

Given the timescale, the fact that no plaque to represent all humankind could ever please *all* humankind, the state of racial and gender politics in 1972, and the gender stereotypes being published in books, magazines and advertisements of the era, it could have been a good deal worse had it not been handled by such a brilliant, committed, well-meaning, humble and thoughtful trio. And the upside is this: the removal of the vulva will remain an illustrative sign of the times in the several-million-year lifespan of the Pioneer plaques.

The team's post-match report, stamped in the NASA archives April 1973, ends: 'It is nevertheless clear that the message can be improved upon; and we hope that future spacecraft launched beyond the solar system will carry such improved messages.'

In the film version of this story, the camera now scans across the Pioneer plaque one last time, the footage slowly fades to black, a needle hits vinyl and through static pops we hear the guitar intro from 'Johnny B. Goode'. Fade up to some archive footage of Carl Sagan striding purposefully. Time for Interstellar Handshake 2.0.

CHAPTER TWO
Needle Hits Groove

'*The Record should be more than a random sampling of Earth's Greatest Hits … We should choose those forms which are to some degree self-explanatory, forms whose rules of structure are evident from even a single example of the form (like fugues and canons, rondos and rounds).*'

Jon Lomberg

The first time I thought about what it might be like to play music to aliens was while watching the feature-length pilot episode of *Buck Rogers in the 25th Century*. I was six. The Voyagers had cleared the asteroid belt and were getting cosy with the gas giants.

For those who don't know it, the set-up for this late-1970s show,[*] a spiritual sister to the original *Battlestar Galactica*, is as follows. An explorer, named Buck, has been sent into space. However, an unforeseen event has blown his ship off course, sending him into a wild orbit that freezes him alive. For decades he drifts, like a bug in amber, before his orbit brings him back to 25th-century Earth. Five hundred years hence, the human race is still going strong, and now enjoys tense diplomatic relations with aliens – aliens that look and sound exactly like us humans, but dress with sprinklings of glam-rock.

In a key scene, the safely defrosted and increasingly confident Buck, played by Gil Gerard, is dancing with a gorgeous Draconian princess. However, it's a stiff and staid shindig, like some futuristic *Pride & Prejudice* ball, and is not to Buck's taste at all. He walks over to the performing musician, waves his hands, gesturing him to stop, like a man flagging down a tiresome street vendor. He asks whether the

---

[*] The character originally appeared in a novella by Philip Francis Nowlan in a 1928 issue of *Amazing Stories*.

gentleman is familiar with 'rock'. He clicks his fingers impatiently to suggest something more upbeat. The poor man, seated behind a banked keyboard and watched by the now silent and scandalised party, comes up with a speedier baroque piece. Buck nips that in the bud quick-sharp, encouraging the musician to let himself go, to 'feel the music'. Suddenly the player discovers funk. He lays down some bass, and Buck purrs with encouragement, nodding his head and duckwalking backwards through bemused and bewildered party guests. With a pelvic thrust, he sashays towards his dance partner, Princess Ardala.

'What are you doing?' she asks.

'It's called gettin' down,' he explains.

And so, with a pinch of all-American 20th-century bravado, the party is transformed by a human time capsule, armed with music from his epoch. Which has some striking parallels to our story.

★★★

Carl Sagan was a magnetic character who loved conversation, a man who could flip between raconteur and attentive listener on a dime, a flirt who loved attention but also backed that up with natural charm and an astounding brain – at least when compared to us normal people. He was good-looking too, with slightly thin, elfin features, a large but not unattractive nose, thick waves of fine brown hair, and an electrifying smile. Then there was his hypnotic, slightly nasal voice, with plenty of treble and base, that seemed to swoop up and down with generous North American 'r's. He was defined by the science he loved, and gave off an abundant and very human enthusiasm for a subject he'd first discovered as a small boy in Brooklyn.

Carl was born in 1934. His parents – Samuel, an immigrant garment-worker, and Rachel – were liberal Reform Jews. In 1939, they took him to the New York World's Fair (where he also witnessed a time capsule being buried at Flushing

Meadows), and he came to view this experience as one of the defining moments of his childhood. Another came in the same year, when he asked his mother what stars were. She told him to visit the library to find out, and when he discovered that our sun *was* a star close up, it fired his imagination.

Evidence of his burgeoning interest in exploration and exobiology can be found in the Library of Congress, which preserves 'The Evolution of Interstellar Space Flight'. This is a collage of imagined newspapers, headlines and stories drawn by the pre-teen Sagan, covering future landmarks in the conquest of our solar system. It imagines our exploration of Jupiter and Pluto, of colonising the moon, a manned mission reaching Mars in 1960, and the discovery of prehistoric-like reptiles on Venus in 1961. And while at Rahway High School (graduating 1951), he entered an essay-writing contest in which he drew parallels between the impact of Europeans on Native Americans and what impact future alien contact might have on humankind.

There's no doubt Sagan courted criticism from his peers for being a careerist, and for daring to make what many classed as populist conjectures about the possibilities of life on other planets, both in our solar system and beyond. His 'talent for popularisation', as JPL director Bruce Murray described it to *The People* in 1980, was bitter fruit to elitists, and yet that was exactly what made him so attractive to the masses. He wasn't afraid to dream or to vocalise his visions. And it wasn't like he was dreaming from a place of ignorance – he was pitching 'what-ifs' and 'how-abouts' from a base of logical scientific probabilities.

Sagan always took great pains to sound notes of caution too, keeping his notions grounded in reality. During a studio Q and A with Patrick Moore on the BBC astronomy stalwart *The Sky At Night*, he was asked about life on other planets, what form life might take, what technology they might have. Sagan immediately pointed out that – while he believed it highly likely, in the many billions of stars and planets and in the infinity of time, that there are civilisations much older than our own, and that these civilisations must have had more time to have developed far beyond our own – it's difficult to say anything useful. He pointed out how human predictions about our own future from just a few decades before are usually wildly off the mark. He talked about Jules Verne, a man given to conjecture, who visualised how advanced travel might look in 1950, imagining a velvet-furnished gondola at the bottom of an enormous balloon.

Sagan reading self-scripted prose on astronomy and the universe would define his public face in America by the time he broke TV documentary records with *Cosmos*.* But for anyone unfamiliar with Sagan, I strongly recommend you watch his talk-show appearances, press briefings and symposiums, as I think he's a man who is at his best when he's playing to a live audience.

In November 1972 he took part in a symposium called 'Life Beyond Earth and the Mind of Man', sponsored by NASA and held at Boston University. He sat alongside figures including professor of physics at MIT Philip Morrison and Nobel Prize-winning Harvard biology professor George Wald. Sagan kicked off with a quote from Scottish philosopher Thomas Carlyle who, when pondering the vastness of space, wrote: 'A sad spectacle. If they be inhabited, what a scope for misery and folly. If they be not inhabited, what a waste of space.' Sagan looks effortlessly cool, and the audience laughs at his deadpan delivery of the Carlyle quote.

They're ankle deep in the debate by now, speculating about what form life on other planets might take, and how humankind might communicate with it. Wald, sounding very Sunday sermon, argues that contact with an alien race, a more advanced race, could be philosophically disastrous for humankind. He talks about how everything we know, as a species, we have learned through hard work, through the toil of figuring it out for ourselves. He argues that being handed more advanced knowledge on a plate could act like a species-wide sucker punch, robbing us of the drive for

---

* First aired in 1980.

discovery, of self-respect. It's a perfectly valid thought, of course. Sagan appears visibly amused. Armed with an incredulous half-smile, he invites the audience to think back. He talks about when he, Carl, was a student learning his trade. He describes how he would go to the library and how, inside that library, there would be lots of textbooks. And inside those textbooks would be lots of knowledge that other people had worked out and written down. 'Now I didn't approach each page going: "Oh my God, they know that also."'

Around this time, Sagan was director* of Cornell University's Laboratory of Planetary Studies. In terms of his career he had endured disappointments (being refused a professorship at Harvard, for example), and was most associated at this point with weather patterns on Venus, the search for life on Mars, and as an articulate speaker on exobiology. Read through editions of the Cornell student newspaper of the day, and you'll soon gain a flavour of his specialisations, interests and obsessions. On 11 September 1972, for example, about a month before that Boston symposium, Sagan spoke on 'Exobiology 1: The Origins of Life'. Later that month there was a report from his light-hearted talk on 'Life Beyond the Solar System' to a packed Statler Auditorium audience.† In March the following year, he spoke on 'Science and Superstition'. In April the *Cornell Daily Sun* reported from a packed panel discussion on science and science fiction – Carl Sagan, Thomas Gold, Sir Fred Hoyle, Andrew Dickson White and Isaac Asimov.

---

* Having started his secondary education at the University of Chicago aged just 16, he worked as an assistant professor at Harvard, before finally becoming a full professor at Cornell in 1970.
† At the Statler gig, Sagan told his audience that 'the likelihood of extraterrestrial life is so high' that an immediate international attempt to communicate with such life is warranted. He also took pains to sound a note of caution that the chance of making contact was very slim.

He'd already appeared on the small screen plenty of times by the early 1970s. Making almost zero eye contact with the camera, he was filmed way back in the early 1960s, discussing the very dim probability of life beneath the 'Clouds of Venus'. Then he appeared as a talking head in the 1966 CBS documentary *UFO: Friend, Foe, or Fantasy*, and of course he was interviewed for TV and radio news for various NASA-related projects, including the Pioneer plaques. Then he took another step up.

In 1973, *The Cosmic Connection* was published, its zeitgeisty title capitalising on the popularity of *The French Connection* and the word 'cosmic' appealing to the counterculture generation of early-1970s America. It was a generously illustrated series of essays, which received the John W. Campbell award for best science book of the year and saw Sagan's name on the bestseller lists. A 1975 Coronet edition included dedications from Isaac Asimov ('A daring view of the Universe by the wittiest, most rational and most clear-thinking astronomer alive today') and Patrick Moore ('Carl Sagan, widely regarded as the leader in his field, is not afraid to speculate'). It also led to the first of multiple appearances on *The Tonight Show* with Johnny Carson, where in December 1973 he was introduced as 'astronomy's most articulate spokesman'. And he had become, for better or worse, a spokesperson for all things extraterrestrial.

He was still a working stiff at Cornell. He would receive the NASA Medal for Exceptional Scientific Achievement for his studies of Mars with Mariner 9. He had chaired US delegations to a US/Russian conference, he was awarded the Prix Galabert, the international astronautical prize. He was also now editing the planetary journal *Icarus*, which he had launched. And let's not forget he was also a working husband and father. By the mid-1970s Carl had three sons: two by his first wife, evolutionary theorist and biologist Lynn Margulis; and Nick, with his second wife Linda. Indeed *The Cosmic Connection* is dedicated to his sons.

★★★

The Grand Tour was approved in 1971, then cancelled the following year, before a cut-back version was green-lit the year after that. By the mid-1970s NASA's primary targets for the mission had become Jupiter and Saturn.

Voyager's principal investigator into low-energy charged particles, Tom Krimigis,[*] describes the meeting where a NASA rep pitched the Grand Tour to President Nixon. With estimated costs in hand, he explained the unique situation, how planetary alignments offered this once-in-a-lifetime opportunity for a moderately thrifty tour of the gas giants. It was pointed out to Nixon that the last time the planets had been so aligned, at the start of the 19th century, President Jefferson had been at Nixon's desk, and he had blown it. Nixon, to give credit where credit's due, said all right but asked that NASA visit just two planets. Fine. NASA had a deal.

Now, imagine you've just told your mum that you're walking to a friend's house down the road. But in your bag you've stowed secret provisions, a couple of apples and a Hershey bar, as unbeknownst to her you are intent on a much longer journey to another friend's house in the next town. It was a little like that.[†] The possibility of directing a ship on to the more remote gas giants, should it survive the encounter with Saturn, remained in the background as a driving ambition – an almost-but-not-quite-secret agenda, designed into the fabric of the mission from the beginning.

Two important JPL figures at this time were Edward Stone and John Casani. In the mid-1970s Ed Stone was Voyager's project scientist. Judging from contemporary photographs in the JPL Archive, Stone often favoured a brown suit jacket with a check design. He looks thin and good humoured – the way scientists look in comic books and Marvel films. He would go on to chair mission press conferences into the 1980s

---

[*] Speaking in Emer Reynolds's 2017 documentary *The Farthest*.
[†] Yes, in this analogy your mum is Richard Nixon.

and '90s, hold the position of JPL director for 10 years, and make regular appearances in 2017 for Voyager's 40th-birthday celebrations.

Casani was born in Philadelphia, studied at a Jesuit high school, entering University of Pennsylvania as a liberal arts major before joining the military. He switched to an electrical engineering major, graduating in 1955, and joined pre-NASA JPL, when it was still administered by the Army Ballistic Missile Agency, having already worked on multiple Mariner and Pioneer missions.

In October 1974[*] the entire JPL Voyager team was conducting a comprehensive Mission and Systems Design Review – in other words a survey of problems that needed solving. The survey covered a lot of ground, from computer command systems, and communications, to how the probes might fare in the extreme environmental conditions to which they were to be subjected (including radiation levels encountered by the Pioneer probes). Then, once a problem was identified, it would be allocated to the appropriate mission team, who then had until March 1975 to come up with a solution in time for a final design review. The October survey highlighted 116 concerns.

At the time of the sitrep, Casani was Chief of Division 34 – JPL's Guidance and Control Division. It was he who, using one of NASA's standard Concern/Action forms, highlighted that there was at present no plan to include a Pioneer-like message with the probes. He added the three-word solution: 'Send a Message!' I've attended meetings where I've raised something trivial just to avoid seeming like a waste of space. I've also deliberately reframed a point that had already been made by someone else, in order to seem like I'm contributing. Now, I'm not suggesting for a second that this was Casani's motivation, or that the problem he raised was trivial. I'm merely trying to point out how, to this layman at least, when put alongside the enormously complex problems that must

---

[*] According to Henry C. Dethloff and Ronald A. Schorn writing in *Voyager's Grand Tour*.

have faced those NASA technicians during the months of design review, the existence or not of a message to the cosmos was like a band member worrying about T-shirts before a song is written.

Casani's suggestion wasn't acted on immediately, of course, but it was recognised as an important consideration. NASA might have had its fingers singed by the reaction to the Pioneers, but there had been a reaction. It had generated publicity, interest. It had excited the world's popular imagination.

The Voyager mission at this moment was known as Mariner-Jupiter-Saturn 1977, or MJS 77. Even with the abbreviations it was a clunky name. Just compare it to Explorer 1, Viking 2, Mariner 1. They sound cool. Mariner-Jupiter-Saturn 1977 sounds like a lumbering triple-vinyl concept album. It's a utilitarian name that encapsulates the plan – that a Mariner-class ship is off to Jupiter and Saturn – but otherwise sounds dull as all hell. Or at least that was the opinion of Casani, who was promoted to become JPL's new project manager in 1976, and immediately began bucking for a more imaginative name.[*]

Unsurprisingly, he met plenty of resistance. Think about it. You're in a band called Gauntlet. With a name like that, you probably play old-school heavy metal. You've built up a following, landed regular support slots at local venues, worked on logos and T-shirts, have plans to book studio time for a first EP any day now. Suddenly your new bassist, who only joined the other day and has only just mastered that tricky middle section in 'Pox Ridden', pipes up with: 'Gauntlet is a

---

[*] Casani was not one to avoid controversy. Despite the climate of concern that budgets might be slashed, despite the fact that Casani's boss, Bud Schurmeier, had impressed on his new charge just how close the whole thing had been to being completely canned and that he did *not* want anyone talking about Uranus, soon after taking his new position at the head of the team Casani mischievously asked for a new telephone number: 4MJSU – Mariner-Jupiter-Saturn-and-Uranus.

rubbish name. Let's call ourselves The Gaunts.' It's really annoying. It's annoying because you've spent a long time working with your current name, you have an emotional attachment to that name. You're invested. You have patches sewn into your jacket, badges on your lapel and a new, just-printed T-shirt. But it's most annoying because deep down you know the upstart bassist is right.

Casani wanted a name that marked this mission out from its predecessors. This was not a mere iteration of the Mariners. Despite the fact that the plan was to save money where possible and to avoid too much tailored engineering by using extant hardware for the new probes, this wasn't Mariner 5.0, or Pioneer 3. It was a pair of brand-new probes going on a trip to the outer planets. This wasn't another *Rocky* sequel. It was *Rambo*.*

Casani was a popular leader. He took over the Voyager project reins from Bruce Murray. Murray was a brilliant man who had pitched hard for the Grand Tour, but he was more scientist than leader, a man who gravitated towards 'push' rather than 'pull' behaviours, more rock than sponge. NASA's director of planetary programmes Robert Kraemer described him as someone 'not known for his timidity'. That's not to say Casani was popular because he was some fluffy pushover. Far from it. In talks, press conferences and public appearances he gave off natural, inclusive charm, built around military-grade steel. Speaking in 2009 to a NASA leadership academy, he summarised what he saw as the essential ingredients for managing successful space

---

* Inconsistent analogy alert. *Rocky II* didn't appear until 1979, and the first *Rambo* film was released in 1982. So the Rocky/Rambo analogy would not have been in Casani's mind back in the mid-1970s. I could have used *Jaws*, as he might have known that *Jaws 2* was on the way (released 1978). And perhaps I could have used *Piranha* as a *Rambo* replacement, as it appeared in the same year. But 'this wasn't a *Jaws* sequel, it was *Piranha*' just didn't seem to work so well. *Death Wish II* wouldn't work either. It came out in 1982.

projects: 'Toughness ... because this is a tough business. It's a very unforgiving business. You can do a thousand things right. But if you don't do everything right, it'll come back and bite you.'

This tough upstart bassist pushed through the name-change resistance – a 'firestorm' according to *Voyager's Grand Tour*. Indeed the MJS team had already gone through a contest to select a working 'MJS 77' emblem for the mission, and the winning emblem had been picked. None of this derailed Casani and in the spring of 1976 they held a second competition, this time for a new name, with a case of champagne going to the winner. MJS had finally become Voyager – although NASA did not officially accept Voyager as the new name for the two spacecraft until March 1977.

So by 1976 the project team had a nice new name in the offing. They had a nice new boss. And they had Casani's three-word call to action: 'Send a message.'

★★★

Between Pioneer and Voyager, both Carl and Frank Drake worked on other space-bound messages. Indeed, if you think of the Pioneer plaques as humankind's first postcard to the stars, Frank was about to send our first email.

Frank was already famous in astronomy circles, not least for his Drake Equation – a formula that summarised all the variables relevant to establishing the number of intelligent civilisations that existed in the galaxy. It was a thought experiment, an attempt to look up at the vast heavens, consider all the factors that made life a possibility, predict how many stars might have planets suitable for harbouring life, and therefore how many civilisations there might be. It was originally conceived by Frank as a tool to stimulate debate at the first meeting of the Search for Extraterrestrial Intelligence (SETI) at Green Bank, West Virginia, in 1961 (the site of the first modern SETI experiment, 'Project Ozma', the year before) rather than as a genuine attempt to come up with a number. According to Keay Davidson's Sagan biography, as Frank spelled out his equation for the first time on the blackboard, the sounds of talking behind him gradually diminished. The 10 attendees (Dana Atchley, Melvin Calvin, Frank Drake, Su-Shu Huang, John C. Lilly, Philip Morrison, James Peter Pearman, Barney Oliver, Carl Sagan, Otto Struve) called themselves 'the Order of the Dolphin' – inspired by Lilly's recent work on dolphin communication.

Frank had been using his Order of the Dolphin buddies as a sounding board, testing out ideas for messages designed so they might be understood by an alien race, looking for some construct that could form a common ground for communication. Ultimately this led to Frank's first interstellar email, sent on 16 November 1974. It left Arecibo in Puerto Rico* at the speed of light, heading towards a cluster of stars that it should reach … ooh, in about 25,000 years. Now, anyone who's read or seen the film adaptation of Carl Sagan's novel *Contact*† will know that radio signals have been leaking out into the cosmos for decades. But Drake's Arecibo message was much more deliberate than

---

* Arecibo Observatory is a part of the National Astronomy and Ionosphere Center. The NAIC is operated by Cornell under a co-operative agreement with the National Science Foundation. The observatory was conceived in 1960 by former Cornell electrical engineering professor William E. Gordon. It was the largest radio telescope in the world – the main reflector dish measures 330m in diameter and covers an area of 18 acres. It has since been superseded by China's Five-hundred-meter Aperture Spherical Telescope, or FAST, which is 195m wider than Arecibo.

† The film opens with a scene that illustrates this very idea. It begins with a shot of our planet from space, accompanied by the sounds of contemporary radio. Then, as the camera gradually pulls further and further away from Earth, we hear a sound essay of humankind's broadcast history in reverse, ending in crackles, static, Morse beeps, then silence as the camera looks back at the Milky Way from beyond our galaxy.

that. Like the Pioneer plaque before it, and with elements of the Voyager message that was to come, it was designed to be decipherable by any being, using the universal language of mathematics.

Writing about it for *National Geographic* on the 40th anniversary, Frank's daughter, science journalist Nadia Drake, described how the message was not publicised beforehand. It was a secret opportunity that had come out of the giant radio telescope receiving an upgrade. It was now home to a million-watt transmitter, and Frank's message was the champagne bottle against the hull, marking the completion of the improvements.

Frank's message had to be simple. Language would be meaningless, so he planned instead to send a picture made up of shaded squares. Imagine the very first computer game you ever played. Or, if you're a bit younger, think about how *Minecraft* looks. Frank's message kind of looked like that, but black and white.

The message was sent around 1p.m. You can listen to it now on YouTube. It starts with a long steady tone, then a long series of alternating warbles, before returning to the steady tone again. Those two-tone warbles indicate to the receiver either a zero or one. So if you knew the correct grid template, you would simply listen to the warbles and it would tell you which squares had to be black, and which should be left blank, to build up a picture. To give the alien receivers a clue about the correct template, the message was constructed from exactly 1,679 characters – a number that is the product of prime numbers 73 and 23. It was thought alien mathematicians should spot that the number could only be made by those two prime numbers, and so would try reconstructing the binary warbles in the correct template shape – a tall rectangular grid of 23 by 73 squares. Or to put it another way, you listen to 1,629 pulses, and build the pictorial message like a stack of Tetris blocks, in 73 lines of 23 squares, giving you a set of images.

Frank packed a lot of information into his three-minute picture. It included numbers one to 10 in binary code, a depiction of DNA, a map of the solar system, four billion written in binary code (the world population at the time), the average height of a human, and the height of the telescope itself. Before it was sent, he wanted to see if someone who had no idea of the message's contents could interpret it correctly. Who did he test it on? You guessed it: Carl. And Carl did pretty well, only stumbling over some of the biochemical data encoded in the message.

Frank's email is still on its way to M13, the Great Cluster in Hercules. And it's important for our story as it shows one thing – that Frank's focus was sending pictures to the cosmos, rather than sound.

Carl, meanwhile, was working on a message to the future. NASA was preparing to launch LAGEOS (Laser Geodynamic Satellite), a satellite designed to measure continental drift. To allow it to make as accurate measurements as possible, it had to be put in a nice high stable orbit. During the design phase, it was realised that as it was to be more or less impervious to factors that cause the gradual decay of other satellites, its estimated lifetime before burning up in the atmosphere was around the eight-million-year mark. So NASA asked Sagan to work on a plaque, this time aimed at our distant human descendants – a greeting card to explain the where, when and why of the satellite. In *Murmurs of Earth* Sagan compares it to a giant golf ball, which the photos show is a pretty perfect summation.

The LAGEOS capsule is much simpler than the Pioneer message. Designed by Sagan, it shows three representations of the continents of the Earth. The first shows them how they were arranged millions of years ago, the second how they appear now, and the third how we expect them to appear many millions of years hence. Just as Drake's pulsar map on Pioneer had served to communicate both time and place to its audience, so this simple map communicated two things: first, it told the future earthling what it was for (to map the changing continents), and secondly, as continents move about an inch per year, it should be possible to calculate roughly when it came from.

The message was etched on stainless steel sheets (about 4 by 7 inches) and installed in the satellite – one at each end of the bar connecting the two hemispheres that make up the golf-ball-shaped whole. The NASA press release, dated 15 April 1976, finishes: 'Whoever is inhabiting Earth in that distant epoch may appreciate a little greeting card from the remote past.'

NASA had again shown its willingness to engage with non-essentials. Yes, the agency wanted to save money. Yes, it wanted to have projects green-lit by Congress. But it also sought to inspire. And it was recognised that these messages and time capsules had great potential for generating column inches, interest and inspiration. With LAGEOS Sagan had successfully delivered another message within a NASA brief, in a very short space of time (like the Pioneer plaque, it was relatively last-minute). So when the call to action finally made it to the top of Casani's in-tray in December 1976, it explains why Sagan was the first name that came to mind.

When Casani called Carl, the conversation didn't go anything like this:

'Hi, Carl.'

'Hi, John.'

'Remember that Pioneer plaque you did?'

'Yup.'

'Well, we need something like that for the Voyagers.'

'Right.'

'There's virtually no cash. And we're short of time.'

'Perfect. Leave it to me.'

'Oh and Carl?'

'Yes?'

'No nudes.'

★Click★

Carl was in Pasadena at the time, taking part in mission operations of the Viking spacecraft on Mars (the second Viking lander had only touched down at Utopia Planitia three months before*). Writing in *Murmurs of Earth*, Sagan describes how this was his first inkling of another 'pleasant and hopeful' plan to send a Pioneer-like message with the Voyagers. Yet I think we can assume that it must have already at least entered Carl's mind that there might be something in the offing.

Nevertheless, time was already short. Launch was months not years away. Voyager 2 was setting off first,[†] in August 1977, and it was already December 1976. It wasn't like Carl could finish it off on a piece of paper on the morning of launch, then gallop down the runway at Cape Canaveral, waving it in his hand.

Perhaps because of this lack of time and money, his initial thoughts were very low-key. He pondered simply reproducing the Pioneer plaques as they were. He considered a simple 'modest extension' of the Pioneers – virtually identical plaques, this time with bonus extras and hidden Easter eggs for any alien collectors out there. In *Murmurs* he gives the example of something in the field of molecular

---

* It's worth remembering that this Mars mission was by far the most exciting ball in play at the time. Sagan was at the centre of a globally significant mission that might finally be able to answer the whole 'life on Mars' thing, and give heft to Sagan as the popular face of exobiology. Viking 1 touched down on 20 July 1976, and Viking 2 on 3 September that same year. Besides taking photographs and various measurements, the landers conducted three biology experiments designed to look for possible signs of life. They revealed a 'self-sterilising' chemical environment, created by ultraviolet radiation and extreme dryness. In other words: bupkis.
[†] Voyager 2 left on 20 August 1977, and Voyager 1 on 5 September 1977. You can imagine this caused newspaper scribes, broadcast journos and NASA PR staffers heaps of confusion. But the reason was simple: although Voyager 1 was launched days after its twin ship, its trajectory was such that it would be the first of the two to reach the outer planets (indeed, it overtook Voyager 2 in December 1977), and would begin the business end of the mission long before the other. So it made sense that the first vessel to reach the outers should be Voyager 1.

biology – perhaps they could encode some message or image into a Voyager plaque that communicated something of the structure of proteins or nucleic acids. But, as Frank Drake describes it, they soon realised this would be a cop-out.

Towards the end of December Carl instigated the great Voyager brainstorm, tapping up friends, colleagues and academic consultants. First on the list, of course, was his Cornell buddy Frank Drake. Then together Frank and Carl quizzed others in the SETI/CETI* posse, and members of the Order of the Dolphin – people used to thinking about interstellar communication. These included Philip Morrison, professor of physics at MIT, Canadian astrophysicist and Harvard prof A.G.W. Cameron, British chemist Leslie Orgel, Hewlett-Packard founder and inventor Bernard Oliver, British philosopher Steven Toulmin, and a great trio of science-fiction writers in Isaac Asimov, Arthur C. Clarke and Robert Heinlein. This roll call of talent wasn't amassed through some NASA-sponsored call to arms printed in the *Washington Post*. No, this was all under the radar. This was Carl and Frank putting their feelers out to people they knew or had worked with. And despite the glaring lack of female input, there's much to be admired in the list.

Some gave their thoughts through informal meetings. Others were remote consultants, quizzed on the telephone or invited to contribute by mail. Carl was open to anything at this time. There were no bad ideas. Indeed, lots of the ideas that came out of this period did make it into the final design of the Voyager message.

Many were quick to point out the obvious: the chances of either Voyager being found were infinitesimally small. It was a capsule bound for space, yet designed to carry meaning to the audience back home, to, as Oliver put it, expand the

---

* SETI stands for Search for Extraterrestrial Intelligence. There is also the branch CETI (Communication with Extraterrestrial Intelligence), which focuses on interstellar messages designed to be understood by another technological civilisation – Frank's Arecibo message being the best-known example.

human spirit.* Heinlein suggested the Voyagers should be
equipped with a tracker, so future generations could easily
find them.† Arthur C. Clarke, on a phone message left for
Sagan from Sri Lanka on 3 January 1977, suggested a message
be included aimed at any future human visitors who might be
considering tampering with the old girl: 'Please leave me
alone. Let me go on to the stars.' Toulmin wanted the message
to communicate that humans were co-operative, communal
beings, rather than individuals; Orgel wanted to show that
Earth carried lots of water; Cameron had the practical
suggestion of painting the plaque with uranium – its rate of
decay would give anyone who found the thing a rough idea
of when it was produced.

During this brainstorm period, one important theme
emerged. The consensus was that, as the ship itself
communicated a great deal about Earth's technology circa
1977, the plaque should concentrate on non-scientific
messages. Art. At first that might seem odd. I can see the
logic in using the universal laws of physics and mathematics
to attempt communication, as these would seem to have the
best chance of being understood. But still, according to
Murmurs, art was on the agenda at this early stage, with Philip
Morrison proposing they send Leonardo da Vinci's 'Vitruvian
Man'.

---

* B.M. Oliver wrote this in a letter, reproduced in *Murmurs*: 'There is
only an infinitesimal chance that the plaque will ever be seen by a
single extraterrestrial, but it will certainly be seen by billions of
terrestrials.'
† Imagine some futuristic open-top bus tour. Blast-off 9a.m.,
Pioneers by lunchtime, swing by the Voyagers around 3p.m., quick
pop to New Horizons, then back to Earth for dinner. The space bus
pulls alongside Voyager 2. Children and parents observe this strange-
looking dormant creature with open mouths, staring through arched
tubular glass while a bored Saturday-job operative with a laser
pointer says: 'And there you can see the Voyager's famous Golden
Record. The brainchild of the great 20th-century astronomer Carl
Sagan … Now on to Voyager 1…'

Then Order of the Dolphin member, inventor and general dude B.M. Oliver upped the ante, piping up with a suggestion that changed the conversation: sound. He proposed that, behind a plaque, there could be a can with magnetic tape. That magnetic tape would be compatible with equipment aboard Voyager, and on that tape they could have a recording of Beethoven's Ninth Symphony. Boom. Bernard M. Oliver, better known as Barney to his friends, had come up with the idea to send music into space.

Carl was immediately entranced. It was an awe-inspiring thought. But magnetic tape? That wouldn't work. It just didn't have the lifespan. Think of all that radiation and gravity, all those cosmic rays and magnetic fields and stuff. Magnetic tape wouldn't like that. Not one bit. Domestic tapes frequently couldn't survive encounters with machines specifically designed to play domestic tapes,* so heaven knows how any kind of tape would get on in the freezing vacuum of space. No, tape was definitely out. They needed something more robust, yet more simple, more future-proof.

Hmm.

---

* My copy of *Hate Songs in E Minor* by Fudge Tunnel had to be sold with my mum's 1989 Mini Cooper because we couldn't get it out of the cassette player.

CHAPTER THREE

Musos v Scientists

> *'It was clear from the beginning that Carl wanted a record that would reflect the whole planet – the music of the whole planet.'*
>
> Ann Druyan

Some friends of mine threw a party in Cardiff in the autumn of 1996. Although I was destined to miss the party because I was stranded in Southampton with a faulty Mini Cooper, I still find myself wondering what it would have been like to attend. Another party I often find myself daydreaming about – one that I missed on account of not being born – took place in New York in the autumn of 1974, just as NASA was conducting its Voyager sitrep and Gerald Ford was giving Nixon an unconditional pardon. Just around then Nora Ephron was planning a party at her New York pad. Yes, that Nora Ephron,* the genius behind *When Harry Met Sally*. Nora's party boasted a pretty impressive guest list that looked set to guarantee plenty of stimulating, cerebral conversation from a host of interesting personalities.

This party is an important stopping point on our narrative. And when Timothy Ferris turned up, Carl Sagan was already there. On the floor. Laughing.

★★★

In 1974 Timothy Ferris was a twenty-something muso and cool as f★@k. When he arrived at Nora Ephron's party he had been the New York bureau chief at *Rolling Stone* magazine for about two years. When I first read that, I pictured a long-haired full-timer, churning out earnest reviews of prog-rock

---

* At the time, Nora was dating Carl Bernstein, one half of the reporting team that broke Watergate.

LPs from an ash-covered typewriter in an open-plan office containing at least one beanbag. I pictured someone with a hessian sack full of acid-laced war stories of 'happenings' at the Fillmore Auditorium or the Avalon Ballroom. And certainly, according to Keay Davidson's interviews with Tim for his 1999 Sagan biography, Ferris was immersed in drug culture and rarely without a pocketful of mescaline hits.

Tim grew up in the Florida Keys. His father was a sometime boxer and tennis pro who had also worked as a newspaper reporter before Tim was born in 1944. 'We were well off when I was tiny but went broke by the time I was in the third grade, whereupon we moved to Deerfield Beach – then as now a place people retreat to after getting knocked around in Miami.'

By now Ferris senior was driving a cement truck and writing fiction on the side to raise money, including the sci-fi novella *The Fifth Assault*, which was published in *Bluebook* magazine. Tim's mother was a flight attendant on DC-3s. Indeed she had been the chief stewardess of American Airlines and for a time the 'face of the airline' – which is how she met Tom Ferris.

Tim got his first telescope in 1956, founding the Key Biscayne Astronomical Association with his friends aged just 13. If you visit Tim's website (timothyferris.com) there's a timeline where his early life is illustrated by newspaper clippings, including one from 1961 showing Tim looking like Elvis crossed with James Dean, wearing an open-neck, short-sleeved white shirt, his full-lipped soulful face topped by matinee-idol hair.

He sidestepped from law into journalism, working first as a reporter for United Press International, then at the *New York Post*, before hotfooting it to *Rolling Stone*. As the New York bureau chief he was writing all sorts – covering politics, international news, music (of course) and science. But while he came from general-interest journalism, his primary interests were music and science.

Tim's science pieces were completely experimental for *Rolling Stone* at the time. 'The first was a one-page piece on cosmology called "How do we know where we are if we've never been anywhere else?" And it got quite a good response.'

'Quite a good response' is Ferris-speak for the fact that Rolling Stones bassist Bill Wyman, who was renting a place in the south of France at the time, reportedly interrupted a party and made everybody listen while he read the entire article aloud. 'When I heard that,' says Ferris, 'I thought maybe there's an audience for this. I find science to be the best subject. Maybe there'll be readers for it.'

That piece was published in March 1973. Another, printed later that summer, is the first extant evidence of any professional interaction between Carl Sagan and Tim Ferris. Having read and enjoyed Carl's *Intelligent Life in the Universe*, Ferris pitched a Sagan interview to his bosses at *Rolling Stone*. The bosses said yes, Tim travelled to Ithaca and, on a snowy January morning in 1973, sat down with Carl at Cornell University's Laboratory for Planetary Studies. The result is a fairly sober Q and A where Carl waxes lyrical on his speciality – exobiology, exploring the possibility and likelihood of life beyond Earth. It's an interesting conversation, not least because they touch on themes that would run through the Voyager project, and indeed would eventually form the core of Sagan's bestselling novel *Contact*. They discuss how radio energy leaking out into space was making Earth a relative beacon in the depths of space, pouring out of three contemporary sources – domestic television, radar defence networks and the high-frequency end of the AM broadcast band. Carl ponders soberly that these are currently the only signs of intelligent life detectable from a distance, before Ferris points out optimistically that a lot of good soul music was being broadcast up at that end of the AM band.

It's a good article. For our story, the point is that Ferris and Sagan were now moving at a steady pace somewhere within the two overlapping Venn-diagram fields marked 'professional acquaintances' and 'friends'.

'I was in New York and we became friendly and we used to just hang out a lot. A lot of the time we spent listening to music,' according to Ferris.

This tended to be almost entirely classical. Tim was an aficionado. He had a wide knowledge of rock and popular music, but if he found himself with an hour to spare to listen to something, he would naturally gravitate towards the string section. If Carl paid him a visit, they'd often put a disc on the turntable, and pretty much without exception, it would be something classical.

'I'm not particularly a jazz fan so we wouldn't listen to jazz. Sometimes maybe I'd put some Dylan on or something. I would occasionally play him a rock thing – put him on the headphones with something sort of "hot". But he had no background in rock and wasn't terribly interested.

'Mostly though in those days you'd put like an entire symphony on the big stereo and you'd actually listen to it. That's not something that's so widely done today. It was less weird back then. And as you might imagine I did have a huge stereo.*

By 1975–76 Tim had left *Rolling Stone* and was working as a full-time freelance writer. He had a New York apartment with an office high amid the treetops, plus a huge stereo and several thousand LPs.

So that's Tim Ferris. A writer. A boy from the Keys, who grew up watching launches from Cape Canaveral and staring at the sky through his first telescope. Now a muso in his late twenties, a classical fan with a more-than-working knowledge of rock music and a huge soft spot for Bob Dylan. And when he arrived at Nora Ephron's party, he wasn't alone. He arrived with his fiancée, another writer named Ann Druyan.

---

* I asked Tim for a more detailed description of his set-up. 'My stereo evolved from an Arvin radio purchased in 1958 … as I recall it then consisted of two Magnepan speakers of three panels each (back then called Magneplanar) standing around five feet tall by six or seven feet wide – they looked like electrostats but actually were dynamic ribbon speakers connected by pure-copper lamp cable (still my preferred method) to a pair of 400-watt Dynaco power amps, an Apt Holman preamp, and an AR turntable employing a lovely Shure phono cartridge. The components other than speakers were housed in a wall-sized set of shelves that I built myself, which also held some 5,000–10,000 LPs.'

While Tim already knew Carl, this was Ann's first time with the great astronomer. She was a liberal, left-leaning New Yorker in her twenties. She was born in Queens in the summer of 1949, the granddaughter of Latvian Orthodox Jews who had come to the US via Sweden during the First World War. Her father, Harry, ran a knitwear firm. Harry and his wife Pearl also had a boy, Ann's older brother Les. She met Tim through a previous boyfriend named Jonathan Cott – another *Rolling Stone* scribe – and the quick version is that, according to Davidson's Sagan biography, Tim had fallen for Ann around the time she looked after him following a bad trip. She was funny, intelligent and beautiful. She was a writer, but the publication of her first novel, *A Famous Broken Heart*, was still three years away.

She entered the room, and there was Carl.

'I didn't know then what I know now,' Ann says. 'I didn't fall in love with him or anything. I just thought, "Wow!"'

They went on to have one of *those* conversations – the kind of conversation we've all had at one time or another – when there's an almost-audible 'click'. You become two parts of a tiny social jigsaw and suddenly there isn't enough time in every minute for you to talk about everything you want to talk about, there's a constant bubble rising in your chest, and words fall out of you. They noticed their similar upbringings and similar parents, they talked about Trotsky, religion and baseball – Ann was able to impress Carl with obscure facts and figures from the sport's early history.

She had not come to the party entirely blind. She had been pre-warned that an 'amazing guy' would be there. 'I remember the exact instant,' she says. Nora had called Ann to tell her about this 'most fabulous man' who would be there too. At first Ann thought Nora must have been talking about Carl Bernstein, but no, she meant a different Carl.

Nora had met Sagan at an editorial meeting at the *Washington Post*. Carl had asked a question about the solar system, Nora had raised her hand and piped up with the right answer, which seemed to impress Carl, who had made her 'feel good about that'. Basking in this triumph, she decided to invite Carl and his wife Linda to the dinner party. And of course she wanted Ann and Tim to come too.

'And I was like "Great!",' says Ann. 'And I remember she was arranging these magazines in this basket. And I remember how she lit up when she talked about him – even now I remember that. I don't remember anything in between that and walking in and seeing that guy lying on the floor and then talking to him about Trotsky and baseball.'

So Tim and Ann, and Carl and Linda began to hook up for occasional double-date social engagements. Tim and Carl would meet for meals, discuss the latest news or chat animatedly at press conferences. Keay Davidson describes how, soon after Viking 2 landed on Mars on 3 September 1976, Tim went to visit Carl at his Pasadena apartment. Although this second lander would only repeat the gloomy no-life-here news delivered by Viking 1 a few weeks before, Sagan did have a photographic panorama of the surface of Mars. Tim describes how they wrapped the photograph partially around themselves, so they could gain some sense of what it would be like to stand right there, on the surface, where Viking stood. Right around this time Carl wrote an introduction for Tim's first book, *The Red Limit*, which would be published to great acclaim in 1977.

There was also a short-lived road test of working together. Carl had been approached to front a new '*Sesame Street* for science' – the brainchild of *Sesame Street* founder Joan Ganz Cooney. Joan thought Carl would be the ideal frontman, and so, in turn, Carl approached Ann and Tim about putting together a treatment. Though that show went nowhere, it was perhaps the reason why Tim Ferris and Ann Druyan would be among the first people Carl would think of calling about the Voyager record. They were good friends, very smart, knew lots about music, art, science and space, and now he had a baseline experience of working with them.

★★★

From the beginning the Voyager message budget was small. Speaking in *The Farthest* Casani describes how, in that initial call to Carl in December 1976, Sagan had said he could bring the project home for around $25,000.* Even with Congress looking over NASA's shoulder, $25,000 wasn't an eye-watering amount. And from this point on, Sagan operated more or less as an independent star pitcher for the project. He couldn't do anything he liked, of course – there would be plenty of rubber stamps and tick boxes in time, as he knew from the Pioneer project – but, for now at least, he wasn't overseen or monitored. He would not be working from an office at JPL or elsewhere in the NASA organisation. He wouldn't be filling in endless consent forms or making reports at daily meetings. He had his okay, his budget and some ideas, he was pumped about the concept of sending music to the stars, and in Frank Drake he had the perfect wingman.†

In late January 1977, Carl and Frank were in Honolulu. They were there for the 149th meeting of the American Astronomical Society, a gathering that included a get-together for the AAS subgroup – the Division for Planetary Science. Frank was staying with Carl and his family at Kawabata Cottage of the Kahala Hilton. It was a large cottage, named after a Japanese Nobel laureate who had stayed there once, and included a large swimming pool. It was here, some time between 16 and 22 January 1977, that Frank first suggested a metal record.

Frank's primary focus was always sending pictures, as evidenced by his Arecibo message. Images could express and communicate sophisticated information efficiently. A simple metal plaque only had space for one or two images at most, and what more could anyone say about humankind with only

---

* In an interview with one of the record team back in 2007, I was given an estimated total cost of the final project coming in at around $18,000.

† Or, to put it another way, Frank Drake had in Carl Sagan the perfect figurehead to absorb all the political flak while he got on with the important business of sending messages to the stars.

two images to play with? So Frank asked himself: how could a metal object of modest size be forced to contain more information? What if you had a plaque, still made of metal, but with grooves and a hole in the middle. It could spin round and play sound. A record. A metal record!

'Sorry to disappoint you,' says Frank, chuckling slightly when I ask him about it.[*] 'There was no "Eureka" moment.' Today Frank views the record idea simply as a logical, natural extension of the Pioneer plaque, an obvious next step when faced with the problem of how to squeeze more information onto a metal surface: 'I thought: "Well, let's do better than we did last time." Because there is very little information content in the Pioneer plaque. And I thought what are the possibilities? Sending any kind of a document, it's not going to work because the document will fade away and not be preserved. What else did we have available to us? The thing that came to mind, that can carry a great deal of information, was a phonograph record.'

Carl immediately saw the value in the idea. Not only could information be physically rendered onto the surface of the record, but that object would also have a much longer potential lifespan than magnetic tape or any other imagined format. The grooves could carry all manner of information too – not just music. A television picture was essentially a collection of signals at different frequencies, a sound. So a television picture could be recorded as a sound, and encoded onto the phonograph record as well.[†] A vista of possibilities had just opened up before them. If the Pioneer message was a single sheet of A4, the Voyager record was a complete volume.

---

[*] On the telephone to me in the spring of 2018.

[†] You may be wondering: 'How?! How do you put photographs on a record? It's a record!' It's a good question, and in due course there will be more detail about how they actually managed it. For now, the important point is that they already knew the theory of how it could be done – a still photograph could be 'filmed' using a video camera, then the resulting video signals could be converted into sound, then that sound could be put on the record. Capeesh?

Talking of single sheets of A4, one handwritten piece of paper survives from Kawabata Cottage in January 1977 and is reproduced in *Murmurs of Earth*. This is the first written plan for the Voyager message. Written by Frank, it is headed 'Proposed MJS Record' (remember Voyager at this point was still going by the old Mariner-Jupiter-Saturn handle). The document maps Frank's proto-playlist for a single one-sided 12″ disc, designed to be played at the traditional LP speed of 33⅓rpm. It's an interesting document for many reasons, not least for illustrating what we already know: that Frank, from the beginning, was very focused on sending both images and sound via the medium of a record.

'I did a pretty quick back-of-the-envelope calculation of how many pictures we could send in a one-hour record,' he tells me. He knew that, in video, 'presently they were sending about 60 pictures a second, but with much greater bandwidth. And I knew the bandwidth for a phonograph record was much smaller – it was at most 15 kilohertz. So I just took 15 kilohertz, divided by the megahertz present in a TV picture, and I realised we'd be able to send about 12 pictures.'

The pictures are often forgotten. From the moment the record went public to today, many people just don't remember the images; it was the music that captured the popular imagination. People might remember 'Johnny B. Goode', but they don't always remember the picture of the wasp. Perhaps it's just a symbol thing: we see a record, we think sound, we don't think pictures. Grab some Average Joes from the late 1970s and ask them what they know about Voyager. The majority won't remember the record at all; out of those who do, only a very small fraction will remember that pictures went with it too. And yet, for Frank, from the outset the pictures were more important than the music.

'We'd already decided the thing to do was send pictures primarily,' he says. 'That was the most important means of delivering information in a definitely decipherable flawless way. We'd had that discussion long ago … What doesn't work is sending any message in a language either written or spoken. The only thing that works when you've had no previous contact, and no ability to send a language course, is what works with babies or foreigners, and that is to use pictures. So we'd already decided pictures was the best way to carry information. It allowed you to carry a great deal of information free of ambiguities and misinterpretations … My thought was that an ordinary television picture was pretty good, so how many bits are there in it, and how many could we send, and the answer that I came up with was 12 or so.'

Frank's first playlist comprises 14 items in total, with an estimated elapsed time noted alongside. At the bottom Frank has written that they were assuming pictures of 500×500 lines, so 250,000 pixels, at 4 bits per pixel. At the time he estimated that each picture might take up to about three minutes of the record's run time.

The tracklist starts with a picture – a spacecraft at launch, with human figures (echoing Frank's Arecibo message). This is followed by more images – human figures ('child, adult, man and woman'), a house, plants, perhaps an automobile. Then Frank suggests sound, specifically a dinner conversation. Voyager playlist 1.0 also shows that they were keen to marry sound and image so that one aided the interpretation of the other. Around 15 minutes in, Frank pitches a photograph of Times Square, closely followed by 'sounds of Times Square'. Then we have a suggested photograph of Sydney Opera House, followed by a symphony, and finally the Taj Mahal, followed by some 'Indian music'.

In some ways the most remarkable thing about this document is that it reveals how relatively modest the first version of the Golden Record was. The gulf between this and the end result is vast. As the project gathered pace, the team would figure out ingenious ways of squeezing more information onto the surface of the record.

★★★

With a plan in hand, Carl left Honolulu pretty excited. This could work. And while the idea of sharing human art, music and information with the cosmos was immediately exciting to all the team, it was music in particular that appealed to both Carl the scientist and Carl the populist. While previous messages might have encapsulated how we think, this would be the first to communicate something of how we feel. Besides, there's a science to music. Carl would frequently and fervently quote a paper by Sebastian von Hoerner, of the National Radio Astronomy Observatory in Green Bank, which argued that the physics of sound offered only a limited number of musical forms, meaning music too could prove to be a universal language, governed by the physics of sound and the mathematics of harmonics.

This interest in the connection between music and the cosmos had cropped up before. For much of his working life, Carl had been in the habit of dictating to his long-serving secretary and personal assistant Shirley Arden. This is the way he wrote letters, speeches, novels and scripts. He spoke aloud, Shirley typed, then Carl reviewed the notes. He also had a slightly Alan Partridge-like method of keeping track of his thoughts in the form of an ongoing 'Ideas Riding' file. These included everything from ideas for novels, non-fiction works and documentaries, to scientific explanations and random thoughts. One typed Ideas Riding page dated 21 June 1973 sees Carl pondering music as a means of interstellar communication. It reveals that he was already thinking about music as a mathematical form, as a more universal language, and whether we would have the wit to distinguish certain types of alien 'music' from noise. Another Ideas Riding[*] file dated 25 March 1975 sees Carl musing that whale song is 'very much' like cello music. Then he envisages placing cellist Mstislav Rostropovich on the poop deck of a Caribbean boat to harmonise with whales.

'He [Carl] knew the pictures were the best way to tell about a civilisation,' Frank says, 'but he got obsessed with the idea that somehow we had to impart our culture and how advanced we were and how sophisticated. He thought

[*] You can see both these examples from the Ideas Riding file via the Library of Congress website: www.loc.gov/item/cosmos000093/.

the best demonstration of all those characteristics was our music. Music gives a measure of the quality of a civilisation, that was what he thought. And so he was obsessed with the idea – and "obsessed" is the right word – he really thought it was vital because it was a way of really demonstrating our intellectual abilities … And so he really got obsessed with the music and that was his baby all through the preparation of the record.'

So where are we? So far we have a leader (Carl), the brains (Frank), a brains trust (the SETI/CETI/Order of the Dolphin posse) and an idea (a metal record). Now it was time to assemble the rest of the team. Sagan began making calls.

Let's imagine Carl sitting at his desk in Cornell, flicking through a Rolodex. He has a piece of paper in front of him, on which he has already written the names Ann Druyan and Timothy Ferris.

'Now let's see,' he murmurs, working through from A to Z. 'One more … I need someone else … Who else do I know? Someone who likes science and … art …?'

His finger stops at 'L'.

★★★

It's the spring of 1972. You're driving out of Toronto along Highway 401 when you see a man walking by the side of the road. He has dark hair, is wearing a backpack and is carrying something in one hand. As you pull past you realise it's a portfolio. His thumb is out. He must be one of those long-haired arty types. You think about stopping. In fact, you make a snap decision and slow the car. You pull into the kerb, leaving the engine running. You watch in the mirror as he jogs towards you, the ungainly portfolio flapping in the breeze. You roll down your window and exchange pleasantries, discuss journeys, and soon he has gratefully climbed into the passenger seat, the portfolio stowed on the back seat.

The man's name, it turns out, is Jon Lomberg. He tells you he is hitch-hiking to Ithaca to meet his hero.

'Who's your hero?' you ask.
'Carl Sagan.'
'Who?'

A few months earlier, Jon wrote Sagan a fan letter that would shape the next few years of his working life. Jon was born in Philadelphia. He didn't study astronomy or art at college. He was an English major, with a heavy emphasis on linguistics and communication. His interest in science – or this particular branch of science – really began in the 1960s. He was 12 or so when his mum gave him a copy of Walter Sullivan's *We Are Not Alone*, first published in 1964. And today, Jon defines himself as someone who was born to make interstellar messages, telling me: 'I was interested in astronomy long before I was interested in being an artist. I always loved art, but in terms of being a painter, a visual artist, that came much later. For me, for as long as I can remember, the most interesting question was: "Is there life on other planets?" And if there is: "How can we communicate with it?"'

In the pages of *We Are Not Alone* Sullivan, the leading science scribe at the *New York Times*, was reporting on the now legendary meeting held at the National Radio Astronomy Observatory in Green Bank, West Virginia. We've touched on this meeting already, but it comes up again because this was essentially Woodstock for a certain branch of astronomers and for exobiology. This is where the topic of extraterrestrial intelligence, and how humankind might be able to communicate with it, was considered seriously by mainstream scientists for the first time. These weren't *Outer Limits* cranks and kooks, these were respected thinkers in their fields. 'It was the first time I heard the names of Carl Sagan and Frank Drake and the other greats in the field,' says Jon. 'And as the years went by I followed up. I read anything else I could see by Carl Sagan or any books that started coming out on the topic.'

This included Sagan's first popular title *Intelligent Life in the Universe*. This was a 1966 collaboration with Soviet astronomer and astrophysicist Iosif Shklovsky. It was essentially a rejigged, rewritten version of Shklovsky's earlier 1962 work, which broadly speaking helped establish the discipline of 'exobiology' as legitimate scientific study. This was the book that moved the whole subject from the fringes to the mainstream. And it not only inspired Tim Ferris to interview Sagan for that article in *Rolling Stone*,[*] it also fired Lomberg's imagination:

'We're talking about all the issues involved in the origin of life elsewhere, the prevalence of it and how we might communicate with it. This was really the first book to debate the subject from a scientific basis. And that book just had an enormous influence on me. A number of pictures that I started doing just after I graduated from college in 1969 were directly inspired by that book.'

Then, when the Pioneer plaques hit the newsstands, Lomberg was entranced. Writing later, he describes how he saw the Pioneer plaques as 'science fiction coming true', even more than when men had walked on the moon. A 'galactic future' had become possible – humans were building interstellar spacecraft and attaching messages. Unable to contain his fan-boy enthusiasm, Jon put pen to paper: 'Basically I wrote Carl a fan letter, saying how wonderful it was and I included some of the art that I had been working on – inspired by *Intelligent Life in the Universe*.'

Me and my sisters Annabel and Kate wrote to *Blue Peter* twice in the mid-1980s. And to *Jim'll Fix It* once. We never heard back from *any* of those rotters. But Carl wrote back to Jon. And he didn't simply send a photograph of himself marked 'All the best, Carl Sagan'; he wrote back with enthusiasm, expressing admiration for Lomberg's work and suggesting that Jon meet him at Toronto airport when he would be making a connection on his way back from Nova Scotia, where he had been watching a solar eclipse as a guest of the Canadian philanthropist Cyrus Eaton.

Carl told Jon the day and approximate time he would arrive at Toronto, but neglected to mention the airline, flight

---

[*] In the introduction for that interview he described *Intelligent Life* as 'one of the most exciting nontechnical science books ever written'.

number, or which city he was flying in from. What should Jon do? Why, he simply needed to wait at the arrivals gate until Carl walked past. There was another problem though: although Jon was a fan, he didn't really know what Carl looked like, and Carl certainly didn't know what Jon looked like.

Jon's solution was pleasingly geeky. He used the Drake equation, written $N = R^{\star} * fp* ne* fl* fi* fc* L$. Carl had discussed this formula at great length in *Intelligent Life in the Universe*, and Jon reasoned that on that particular day in the Toronto International Airport Carl – and only Carl – would be able to recognise and understand it. Jon wrote the equation in black magic marker on a big piece of paper, taped it to the outside of his portfolio, and stood in the airport.

I know what you're thinking. Why didn't he just write 'Carl Sagan' on the side of his portfolio? Or for that matter, 'Jon Lomberg'. Well, this way is much more fun. Anyway, back to the airport.

'I wandered around the gates as planes arrived,' he writes.[*] 'Many people eyed me suspiciously, wondering what cult I was hawking, until a tall, dark-haired man came towards me with a big grin and outstretched hand saying, "Hi, I'm Carl."'

They talked for two hours. They talked about astronomy, science fiction, art, and the *Encyclopedia Galactica*. Then Carl had to catch another plane back to Ithaca.

'Look,' Carl said, 'I've just signed a contract with Doubleday to write a book for a popular audience. Would you like to illustrate it? Yes? Good! Can you come down to Ithaca within the next few weeks and we'll talk about it?'

And that's where we join Jon, hitchhiking his way down Highway 401 East from Toronto past Kingston Ontario. After you dropped him off, Jon turned south, crossed the

---

[*] Before interviewing Jon in 2017, he was kind enough to share with me an unpublished manuscript in which he described his work with Carl, and his work on the Voyager Golden Record. Henceforth if you read 'Jon says', it's from an interview, and 'Jon writes' is from the manuscript.

border into New York State just above Watertown, down past Syracuse and Cortland towards Ithaca.

'It's a beautiful route that takes you past the Thousand Islands and the Finger Lakes,' he writes. 'I was to drive it many times over the next 20 years, but that first time down I hitchhiked my way across the border into the US, carrying a backpack and portfolio, and telephoned Carl when I got to Ithaca.'

Carl picked him up in an orange Porsche 914. The licence plate read 'Phobos',[*] named after one of the moons of Mars.

Their first work together was *The Cosmic Connection* completed in 1973, a series of essays on various subjects, liberally illustrated throughout, including a baker's dozen artworks by Jon. Later that year Carl called Jon again, this time for a television show about space that never got made. This doomed David Wolper production nevertheless proved to be a training ground, a place where they learned all sorts of lessons about TV production that would come in very useful when *Cosmos* swung into production some five years later. Jon was hired as creative consultant and, before the plug was pulled, was sent off on fact-finding missions to the High Altitude Observatory in Boulder, the National Air and Space Museum (where he quizzed Fred Durant on astronomical art), and Haystack Observatory (where he watched Dr M.L. Meeks running what were then jaw-dropping computer graphics in the form of simple black-and-white vector animations).

Now we come to March 1977, where we find Jon working as a freelance radio producer for the Canadian Broadcasting Corporation.[†] At the time he was recording Carl reading

---

[*] Carl liked to refer to the hurtling moons of 'Barsoom' – the Martian name for their homeworld in Edgar Rice Burroughs's 'Jon Carter' stories. The books thrilled Carl as a boy, and helped inspire his interest in Mars.

[†] He had worked there since 1973, as a writer/producer of radio documentaries for CBC series *Ideas*, covering space science, psychology and meteorology.

excerpts from his latest book, *The Dragons of Eden*, which had just won the Pulitzer, and Jon had recently installed a housewarming gift of a large 3D Plexiglas galaxy in the living room of Carl's new home on Tyler Road, Ithaca.

Jon writes: 'Dotted through the blue stars of the galactic disk were a few bright magenta dots marking the home stars of advanced civilisations. Small panels bore writing in glowing characters, "legends" that described the location, biology, and culture of these imaginary societies … By the light of the galaxy in Carl's living room, he told me about the Voyager Record project.'

Let's imagine Carl looking up, with a faraway look on his face, as he says: 'We think we may also be able to send some images. Between the sounds and the pictures, we can describe something of life on our planet and human emotion. I am assembling a small team to make this record. You could be of great help. Are you interested?'

★★★

The Voyager interstellar message team was now complete: Carl Sagan and Frank Drake, supported by Ann Druyan, Tim Ferris, Jon Lomberg and Linda Salzman Sagan. Throughout February and March the project plan began to expand on the 14-point primer penned by Frank in Honolulu. The record was still designed to spin at 33⅓rpm, but by now they had decided that it could be double-sided. Now, rather than running images and sounds together on a single-sided disc, the music would be on one side, the images on the other[*]

---

[*] To this layman at least, this seems quite a good idea. Do you remember the sound old dial-up modems used to make? Or the sound cassettes made when uploading games to your old Atari? Now imagine you're an alien. You've finally figured out how to play the record, you've sat down with your mates in some alien lab, you've listened to some Beethoven, then a clunking, high-pitched, metallic, gibberish sound comes from this ancient alien artefact. You might not think: 'Oh, don't worry, it's just an image.' But you might think: 'Well, this is a little too avant-garde for me. Skip it!'

(along with other non-musical information). So while total runtime had doubled from Honolulu, one side of music still only gave them around 27 minutes to play with. And, as Carl noted in *Murmurs of Earth*, that's barely enough for a couple of movements of the average symphony.

Carl contacted musicologists and academics. Contributors at this stage included Robert E. Brown (executive director for the Centre for World Music in Berkeley), ethnomusicologist Alan Lomax and a friend of Carl's named Murry Sidlin (at the time conductor of the National Symphony Orchestra in Washington). All three would exert an influence over the direction of the project.

Sidlin was born in Baltimore in 1940 and had studied music at the Peabody Institute, graduating in 1968. His opinion was that they had to include complete pieces – no excerpts, highlights or fragments. One Sidlin suggestion was that they include Stravinsky's *The Rite of Spring*, followed by No. 1: Prelude and Fugue in C from Book 2 of Bach's *Well-Tempered Clavier*. Carl writes in *Murmurs* how Sidlin felt the contrast in mood between these two pieces would be 'striking', and indeed that transition, exactly as suggested, would be adopted come the final mastering of the mixtape.

Lomax is credited with helping spark the global folk and blues revivals of the 1950s and '60s.[*] He started his career by literally following in the footsteps of his father, John, a folklorist who in the early 1930s began a decade-long series of road trips, making field recordings for the Archive of American Folk Song of the Library of Congress. The whole Lomax family assisted in this endeavour, including 18-year-old Alan who in 1933 accompanied his father on the first of many field trips and became the project's first paid assistant.

When funding for the Library of Congress recordings ceased, Alan went freelance, travelling America, Europe and the world, making thousands of recordings, reams of notes and hours of oral histories, preserving music from ethnic

---

[*] Alongside Harry Smith's 1952 six-LP *Anthology of American Folk Music*.

groups, fading dialects that would otherwise have completely disappeared from the historical record, and documenting obscure forms of music. All this sat alongside his recordings of celebrated folk, jazz and blues performers such as Woody Guthrie, Lead Belly and Jelly Roll Morton.

With the FBI investigating supposed communist leanings, Lomax moved to Europe. Then he returned to New York in triumphant fashion, organising the Folksong '59 show at Carnegie Hall. This featured bluesmen like Muddy Waters and Memphis Slim, alongside bluegrass acts and folk revivalists such as Pete Seeger. The audience went wild and the resulting LP (*Folksong Festival at Carnegie Hall*) would influence a new generation of musicians.

In the spring of 1977, when the Voyager team came knocking, Alan had been completely immersed in music from all over the globe for his entire working life and had only just returned from another extensive trip overseas. He quickly drew up a list of music he felt should be included. Some were his own recordings, others were commercially available. Throughout he would campaign vigorously for ethnic or world music at the expense of Western classical pieces. He also wanted them to choose pieces that best illustrated his theories.

During his long career, Lomax had come up with a complex series of classifications for different types of vocal music, which together he called cantometrics.* He believed these measurable forms were found within all types of folk music and could be shown to correlate with certain societal developments. For instance, a certain type of rhythm when combined with a certain type of harmony might indicate a

---

* In cantometrics Lomax developed thirty-seven 'style factors' to categorise forms of folk music – these were produced in consultation with various specialists in linguistics and other studies. So in terms of vocal performance, for example, he would grade a particular 'style factor' (such as level of cohesion in group singing, or length of phrasing) on a five-point scale.

sedentary agricultural community. That's a hugely simplified description, but that's the basic idea.

Lomax was as enthused as anyone about the record. And he felt if they picked the right pieces of music – tracks that fitted into and illustrated these evolutionary theories of musical development – it would be possible to paint a complete picture of human history, of human progress, in the Voyager mixtape. He later felt that Sagan was rather dismissive of his theories.* Reading between the lines, I think Lomax may have been *too* keen, almost too much of an expert. He was someone who could be quick to perceive criticism from others. And it's possible to see how someone could, rightly or wrongly, interpret Sagan's tone as snooty. Sagan admits in *Murmurs* that they simply didn't have the time or space to listen to and incorporate all of Lomax's ideas on cantometrics into the record.

What is not in doubt is that Lomax would suggest several of the most memorable tunes on the Golden Record, on one occasion literally flinging a disc across the room in his enthusiasm. His influence on the project has been overstated at times, while at others it has been unfairly overlooked. In any event, both he and Robert Brown gave significant time, thought and energy to Voyager. And while Alan's relationship with the Voyager record was destined to sour over time, Robert's refusal to give ground on one particular subject would lead to a most goosebump-raising three minutes of gold.

While Jon Lomberg had already been earmarked for Frank's picture team, he too had opinions on what music should be in the running. He writes that when he suggested Mozart during a conversation with Carl, he was shocked at the astronomer's response.

'It's kind of lightweight, isn't it?' said Carl.

'Lightweight?' gasped Jon. 'What are you talking about? You just have to include Mozart.'

---

* See eamusic.dartmouth.edu/~larry/published_articles/voyager_.pdf.

Writing about it later, Jon put Carl's reluctance down to him being a child of the post-war era, when the 'big gooey' piano concertos of Rachmaninoff and the like were most popular. Indeed, Carl's favourite piece of music of all time, he once informed Jon, was the *Russian Easter Festival Overture* by Rimsky-Korsakov. As far as Jon was concerned, they'd be absolutely crucified if Mozart wasn't on the record. Carl still looked doubtful.

'Do you really think so? Well, why don't you put some suggestions on tape, and we'll listen to it.'

Back in Toronto Jon made a tape. Knowing any lengthy piece would immediately be met with stiff opposition, he put forward three 'brief but perfect jewels'. These were 'Voi Che Sapete' from *Figaro*, the 'Alleluia' from the 'Exsultate, Jubilate, K.165', and the 'Queen of the Night' aria from the second act of *The Magic Flute*, whose high F is the highest note sung in all opera. Jon sent the cassette to Carl 'with a little prayer'.

By their next meeting Carl was convinced and had 'Queen of the Night' earmarked for eternity. He approved of the elegant economy of representing, in less than three minutes of runtime, 1) Mozart, 2) the range of a human soprano, and 3) the category of opera.

'We'll send "Queen of the Night",' he told Jon.

Mozart was aboard.

***

This variety of music pouring in was all very well, but the record was still painfully short. They would soon have suggestions enough to fill 10 records to the brim with good music, and yet only one side of a single LP to play with. It was so unsatisfying. With so little space, how could they please anyone? Put yourself in their shoes. Give yourself just 27 minutes to show off your own collection and see how far you get.

There were lots of fairly lengthy classical pieces in the mix too, all jostling for position. Carl was in team Debussy, Tim had the back of Bach's 'Passacaglia and Fugue', while Lomax was pushing a Sicilian folk song. Tim, like Murry Sidlin, was dead against using any excerpts or samples – there was to be no fading in or out on *this* record, thank you very much; complete pieces or nothing – and the 'Passacaglia and Fugue' was just too long.

They began to talk about ways they might increase the records' capacity. Sagan writes about how he and Tim discussed various ideas. They pondered making it a double album – sending two pairs of records bonded together, making four sides in total, which could double the playing time for pictures and music – but NASA was working out all its calculations on the assumption of the weight of one single record mounted on the side. Sagan knew that if he went back to NASA and asked to move the goalposts, it was bound to cause trouble, and probably get a big fat no.

Frank explains: 'They had allowed a certain amount of weight for the record … In a nutshell, on the spacecraft there are little rockets that are fired to orient it, and to change its speed slightly. And those rockets, their thrust vectors have to go through the centre of gravity of the spacecraft. And any time you change any object on the spacecraft the position of the centre of gravity changes and you have to re-orient all these things. NASA was at the point where they didn't want to change anything – add any weight or move things around – because they'd have to re-do the whole thing, and if you get it wrong the mission fails because when you fire the rockets, the spacecraft, instead of doing what it's supposed to, spins. It's bad, bad news. So NASA was very nervous about all this and wanted to make sure we didn't do anything which would require us to change the weight of the record.'

So they talked over narrowing the grooves, pushing them closer together, but this might leave the disc unplayable, and anyway would result in quite a heavy loss of sound quality.

Sound quality? 'Hang on,' thought Tim. 'What about talking books?'

Record players made today normally come with the 33⅓rpm or 45rpm speed options. Older machines – indeed many of those manufactured during the 1960s and '70s – still came with the 78rpm speed for playing shellac discs, and many also came with a 16rpm option.

At 16rpm (or, more accurately, 16⅔rpm) records span too slowly for proper high fidelity. However, the slower speed massively increased the record's capacity, making this a natural format for longer, spoken-word albums. In reality, the 16s never dominated record sales in the way that 33s, 45s and 78s had before them, but the 16rpm format was used for talking books for the blind. Radio stations also sometimes used 16rpm discs to make pre-recorded broadcasts, and there were all sorts of curios – such as the legendary Seeburg 1000 record player, designed to play 9-inch 16rpm discs for background muzak in public spaces, where sound quality wasn't paramount. However, with the onset of the more compact cassette from the 1960s – ideal for talking books and longer, spoken-word albums – the 16rpm format was doomed.

Tim says: 'We were starting to pull things together. And it seemed clear to me that we weren't going to be able to say very much in such a short amount of music. And I just thought of it one evening; it wasn't a big discovery or anything, I just thought: "Oh hell, we can do the talking book thing." I looked up the stats and the roll-off wasn't that bad, we weren't going to get big high-fidelity issues or anything. Compression wasn't a problem. It just seemed like an obvious solution, so we did it without any hesitation.'[*]

Earth had 90 minutes.

---

[*] Tim charts the specific outcome of cutting the record at 16⅔rpm in *Murmurs*. He also observes that an alien listener might note that some music on the record is in stereo, while other tracks are mono. This might lead the listener to interpret that some of the music made on Earth was composed by a creature with only one ear.

CHAPTER FOUR

Uranium Clock

> *'It was an absurd task right from the start. Describe Earth and Humanity in 100 pictures. Do it in a few weeks.'*
>
> Jon Lomberg

The biggest obstacle was time. The completed artefact had to be delivered no later than 1 June and there were suddenly an awful lot of moving parts. The expanded plan for the golden record now included images, music, a sound essay and some kind of spoken greeting. But with a little over six weeks until their unmissable deadline, Carl needed to delegate.

By late April the various teams and sub-teams had taken shape. Frank took on all technical aspects of the project, recruiting physicists Valentin Boriakoff and George Helou to help. Cornell's astronomy department provided further back-up in the form of engineer Dan Mitler, staff photographer Hermann 'Eck' Eckelmann and NAIC draftsman Barbara Boettcher. The project would generate a huge amount of tedious legwork too, in terms of copyright clearances, letters, phone calls and all the other bits of legal admin, much of which would be shouldered by Amahl Shakhashiri (Frank's assistant at the National Astronomy and Ionosphere Center), Wendy Gradison (Carl's editorial assistant at *ICARUS*, the International Journal of Solar System Studies, which he edited) and Shirley Arden (Carl's executive assistant).

Tim and Ann worked mainly from New York, while Carl, Frank and Linda generally stuck to Ithaca. Jon Lomberg commuted between Toronto and Ithaca, staying on Cornell campus during the final push. Jon writes: 'It could have taken years, but we didn't have years … And I don't know if we could have done it any better, if we had years to do it.'

The nature of the project was organic, with an expanding brief, no rigid remits, and a feverish pace. In May and June 1977 there was a lot going on, and for the next few chapters of our story this is going to be reflected by a certain amount of hopping about.

***

You'll recall that Frank's initial 14-point plan included sounds of Times Square among the suggested audio tracks. As the record concept took shape, so this aspect of the project expanded, eventually becoming a distinct chapter within the final record – a sample of Earth's natural and human-made sounds.

Jon had already lobbied hard for Mozart, as we know, and would submit a full hour-long tracklist in time. But now Carl asked him to send over some ideas for the proposed sound essay. Jon was working on documentary pieces for Canadian public radio at this time, which essentially involved pulling together sounds to create audio pictures. Now he had a chance to pitch a sound collage designed for aliens. He conferred with his colleague, CBC radio producer Max Allen, and they began to draw up a plan.

It seemed logical to Jon and Max that they should order the sounds in a broadly evolutionary sequence, reflecting the evolution of life on Earth. Jon suggested starting with natural sounds, then sounds of non-human life, then human life, then sounds of human society and the urban environment. They could, he wrote in a letter to Carl, decide to arrange sounds by frequency or tone, grouping similar sets of sounds together. He reasoned too that certain sounds – such as that of weather systems or of water flowing over rocks – wouldn't necessarily be as foreign to an alien as sounds of human speech. Jon wrote out his completed montage plan* and submitted it to Carl in early April.

---

* Which bears a striking resemblance to the final record's sound essay.

In early May 1977, Ann and Tim visited Linda and Carl at their house in Ithaca. It was a bright spring day, and the group gathered around the dining table and began to brainstorm, writing down sounds that might suit an audio essay of life on Earth. Wendy Gradison was there too, a recent college graduate who had followed a Williams College boyfriend to Cornell and by this point was working as Carl's editorial assistant. She found Carl to be 'fabulous ... Attractive, pleasant, full of energy, kind, creative and dynamic.' Wendy is a broad-smiling bundle of energy herself, warmly remembered by everyone who worked with her. She first heard about the project while sitting in her shared office with astrophysicist Dr Steven Soter,* right across the hall from Carl's office. There she was approached by Shirley Arden, Carl's executive assistant.

Wendy says: 'It was a gift to me that I was able to be in the orbit of someone who was not mainstream, was not religious in the traditional sense – which helped me develop my confidence to believe what I believed to be true, or not, and act as I saw fit to act. To thine own self be true and all of that ... I was young, impressionable and lucky.'

Shirley is another of the forgotten heroes of the Golden Record. She was in her forties and had been working for Carl since 1974, usually based at Cornell, but also accompanying him to his temporary office at JPL. By all accounts she was efficient, unflappable and a very fast typist, who was at the centre of all Carl's activities during some of the busiest, most successful and productive years of his life, handling travel arrangements, expenses, contracts, letters, writing and lots more besides. A *Cornell Chronicle* profile from September 1976 describes Shirley's working days during the Mars Viking landings, starting at 2.30a.m. fielding calls between Carl, the lander imaging team, and correspondents from ABC or the

---

* LA-born Soter achieved his PhD in Astronomy at Cornell in 1971. Between 1973 and 1979 he was working as assistant editor of *ICARUS*, and would go on to co-write Sagan's documentary *Cosmos* and the eventual 2014 follow-up helmed by Neil deGrasse Tyson.

BBC. Some 20 years previously she had worked in the UN office of Swedish diplomat Dag Hammarskjöld (who died in a plane crash in 1961), before taking a break to become a full-time mum, returning to work after her son Erik started college. She first applied for the job with Carl because her daughter Jenny had read *The Cosmic Connection* and urged her to go for it.

Anyway, back to the Sagans' dining-room table and the brainstorm: birds, bees, streams, surf, wind, thunder, lightning, crowds, traffic, trains, rockets, barks, children playing ... The guiding principle was that the sounds should communicate something about the story of our planet and human activity on it.

Ann and Tim left that meeting with a notebook of ideas, and lots of work to do. Tim's role was already moving towards the production side of things, so it was Ann who would undertake the immediate legwork of compiling sounds. The following day Ann was back at her place on West 74th Street, where she hit the phones. She called sound libraries, universities and archivists across the country, looking for the best recorded examples of each sound: 'In this little voice I would say: "Um, hi ... Um. We're creating an interstellar message for NASA." And people immediately thought I was out of my mind! "And we'd like to have your *this* or *that* for an interstellar..." People would hang up!'

She reeled off the same story again and again: 'Yes, NASA is making a record. Yes, NASA is sending the record into space. No, this isn't a hoax. No, I'm not selling anything.' She was met with positivity, enthusiasm and help, diluted by some push-back. Many of the sound libraries opted out through simple confusion, others through mistrust of any government-sponsored project. To be fair, it was a strange request. Even today, if you tell a person about the Voyager record – someone who's not heard of it before – it excites head–scratching, furrowed brows and scepticism. People think you're making it up or that you've misunderstood something. So imagine what it must have been like in 1977.

Cash was a problem too. With hindsight we can marvel at this short-sighted attitude – sound libraries missing out on a shot at immortality just because of a few dollars and cents. Nevertheless these were commercial businesses and, unsurprisingly, many didn't care what the project was about, they just wanted to be paid. Not only that, they wanted to be really rather *well* paid. But as this was being done on a relative shoestring, Ann couldn't really offer them much more than the cost of the tapes that the sounds would be provided on, and the cost of shipping. However, she received lots of very enthusiastic responses. Mickey Kapp – then-president of Warner Special Products – put the entire Elektra Sound Archives at Ann's disposal; Dr Roger Payne of Rockefeller University was delighted to help with her request for whale song; and Alan Botto of Princeton would end up supplying the sounds of a freight train and of a rocket launch – in fact a recording of the Saturn V lift-off from inside mission control.

Yet there was some hostility too. Ann writes in *Murmurs* about one particular individual she called on in person because she had heard he could boast an unrivalled collection of recordings of children's street cries. He threw her out of his office, saying NASA had some nerve sending a 'little girl' to talk to a 'big soundman' like him.

One of my great enthusiasms for the Golden Record is the way it seems to have a mischievous personality of its own. It bites you when you're not looking – like a cat that's suddenly decided it's had enough of this 'petting' business – for the discs managed to leave us with a fuller, more warts-and-all picture of humanity than even the Golden Record team intended. The discs contain secrets, errors, odd patches of darkness. And what I particularly love about our 'big soundman', who in my mind wears a white singlet and chews a cigar, is that his idiotic misogyny will last for a billion years. All right, his echoes aren't on the actual physical record – the aliens won't know about him – but as long as we are around to discuss the history of the making of the record, he will be remembered. For the rest of human history he will be an unhelpful asshat. An amazing thought.

Ann wasn't deterred. She was after the definitive examples of each sound. She wanted the froggiest frogs, the doggiest dogs, the perfect kiss, the angriest hyenas, the most destructive-sounding earthquakes, proper full-blooded chimps, more dogs (they wanted wild and tame ones, you see), sheep, Morse code, hammering blacksmiths…

That all might sound pretty straightforward on the surface, but Ann was working without a computer. She had a phone, a pen and a notebook; there wasn't any Google or YouTube to browse videos or sound files. Besides, if you've ever tried to track down the definitive *anything*, you'll know it can prove surprisingly hard. If you've ever done any picture research for a magazine, blog, website, profile or whatever, you'll know how it feels. You have an image in mind, an image you feel you've seen a million times in a million places, and then when you come to actually pick one that fits the bill, you just can't find it. Before you know it, you're using something that's not quite right. And while I've never compiled a sound essay, I assume the same problems arise. 'Yes, yes. That's a very nice sounding frog. Have you got anything a little more … froggy?'

★★★

Robert Brown wrote to Carl on Monday. He had heard from Carl the previous Thursday (5 May 1977 – the same day Ann, Tim and Wendy visited his house in Ithaca), when Carl had told him the good news that the record's capacity was to be extended. So in his 9 May letter, Robert includes suggestions for 38 minutes of music, mainly picked from commercially available recordings, and emanating from both Western and non-Western traditions.* He attempts to include music that is

---

* He argues that any such list, seeking to capture so much in such a short space of time, could never be well decided by committee. No two ethnomusicologists would ever compile the same list. And he compares the problem to that of a botanist asked to represent the planet's flowers with just six varieties. His entire letter is reproduced in full in the appendices of *Murmurs of Earth*.

simple, and pieces that are complex. He chooses music that captures various pitches, types and timbres of the human voice, and music with a wide variety of meter, tempo and rhythm, and scale and harmonies. He also attempts to choose music from a range of time periods, to trace a kind of shorthand evolution of human music – partially mirroring Lomax's ideas.

In this proposed playlist, he has seven pieces, each with an LP catalogue number and a detailed explanation about why he feels it would be a valuable addition. His first pick is 'Indian Vocal Music' by Kesarbai Kerkar (catalogue number: HMV EALP 1278). It's three minutes and 25 seconds long, he tells Carl, is a solo voice, with a seven-tone modal melody with auxiliary pitches. He notes that it has a cyclic meter of 14 beats, alongside drone, 'ornamentation' and drum accompaniment and some improvisation. He also gives a partial translation to the words of the music: 'Where are you going? Don't go alone…'

Each suggestion comes with similar explanations about the musical form it represents. There's a Javanese gamelan, which was his particular area of expertise, some Debussy (varieties of instruments, contrasts of dynamics, complex harmonies) and John Coltrane's *Giant Steps*. Brown then lists a host of other types and tracks that he would include if they could: a lively mridangam solo in a tala of five beats, a West African dance piece, an unnamed example of 'electronic music'.

The letter shows the kinds of conversations Carl was having, and the approach and ambition of one of his key advisers. It shows the pair had discussed individual potential tracks beforehand. Specifically Brown describes how a trip to record shops in Portland had failed to produce Carl's 'Chavez piece'. Most of all, Brown's letter shows a man who is clearly excited. It ends with a rather sweet paragraph in which he thanks Sagan for what he is doing. He talks about how it has inspired him to look again at his own work, how pleased he is that someone is doing something that pulls together astronomy and music, something he had explored in a paper a year before, but had not detected any enthusiasm for. He talks about how working on the project – however it turns out – has given him a jolt of energy, has swept away some of his 'mental debris', and helped him focus on his life's work.

★★★

The Voyager record was being designed for two audiences – human and alien – and, as it was so unlikely to ever be found by aliens, it was much more akin to one of those time capsules (like the one Carl witnessed being buried at the 1939 New York World's Fair) than a message in a bottle. That's as maybe. For Jon Lomberg, who pitched up at Cornell in early May, it was all about the aliens. He had already been involved remotely, as we know, but now he was in town, sleeves rolled up, rubbing his hands together, ready to get stuck in.

Within their six-week schedule, they had to allow time for NASA to see and approve the contents ahead of final mastering and production. That gave them more like four weeks to make all the selections, figure out all the technical problems, clear copyrights and permissions and other legal whatsits, before showing NASA what they had. And all along, the entire team – from Carl and Frank on down – were toiling under the constant threat that NASA might simply say no at the final review stage.

When Jon arrived at Cornell they didn't have a single image selected. They had plenty of ideas, themes and subjects, and he had plenty more of his own, but not one image was on the table. Jon would be working most closely with Frank.

Jon says: 'I describe Frank as a kind of cross between Albert Einstein and Thomas Edison in the sense that he's theoretically brilliant but he's also very much a man of his hands who likes to build things. He's a craftsman, he cuts and polishes gemstones and … a remarkable man. Carl was the figurehead – and he was the leader of it, and he plotted the grand strategy – but in terms of the technical brilliance, it was all Frank.'

The number of images they could fit on the record was increasing. When Carl first talked to Jon they thought maybe 12. By early May, it appeared they'd have room for a lot more – in the region of 100. This depended on how many colour images they chose to send. Although the exact timings of sound-converted images were not yet certain, they knew that black-and-white pictures would take up about a third of the space of their colour counterparts.

From the Voyager brainstorm, certain ideas had taken hold. The first is the most obvious: they wanted to communicate as much information as possible about Earth. They also decided to sidestep politics and religion. Avoiding images with overtly religious meaning might have been a harder step to take in today's climate, but operating without constant NASA oversight meant such a decision could be made quietly. It's logical too if we examine it from the perspective of an alien audience: any image showing religious iconography, or some religious figure or structure, is unlikely to carry much weight in terms of its interpretation by an alien being; instead, any mysterious iconography would add another needless layer for an outside observer to interpret. Plus, if they had chosen to show a mosque, for example, they would have then felt obligated to also include a cathedral, a synagogue and so on. (Indeed, much later a woman in Italy chided Jon while he was there on holiday for not including a picture of the Virgin Mary. He tried to explain that if they had showed a symbol from one religion, they would have had to show symbols from all.) And these would have only been included to satisfy an Earth-based audience, rather than an alien one.

Jon says: 'I think that Carl – and perhaps some of the others, although I can't say for sure – was more aware that this project would have potential importance back on Earth. So I think he was more concerned with the human audience. And that meant making sure the message was global in feel – it wasn't coming from Americans, or from developed Westerners, it was coming from the whole world – and that we didn't bias it by putting in our religion and our philosophy. Of course, for the aliens that wouldn't make any difference, but for people on Earth that would give them a much greater sense of inclusion, which is what he wanted to achieve.

'I was much more interested in the extraterrestrial audience. I really wasn't thinking of the human audience very much at all. I would say the ratio was maybe 95 per cent to 5 per cent in terms of my attention.'

There was no room for pictures of mass protests or influential leaders either. And they didn't want to show any image that might seem threatening or hostile. So, for instance, a picture of a nuclear explosion – which would communicate something of our technological expertise – was ruled out as it might be seen to be a threat.

Most importantly of all, they had to try to choose images that could be correctly interpreted. Remember the old Pioneer plaques? Back then they had thought of having the human figures holding hands until they realised that an alien being could interpret the two figures as a single organism. That kind of consideration was there throughout the image hunt. Explaining this to the local press around launch time, Frank used the example of jewellery: how to a human audience the use of jewellery is obvious, but how another civilisation might assume it to be part of the human anatomy or some kind of identifying mark.

Tim and Ann, meanwhile, travelled to Washington DC for a few days. During the day they visited the National Geographic Society's sound collection and the Library of Congress. In the evenings they took part in a series of late-night debates with whoever from the record team was in town.

There was little that was controversial in the excerpts being mulled over for the sound essay: birdsong, engine noises, waves, animals, wind. Then at the Library of Congress a sound engineer arrived with a shopping cart full of records in various sizes, speeds and formats. He said they couldn't touch the records, they had to point and request. One sound that briefly came under consideration was a very early field recording of battle taken from the First World War. In it an American soldier could be heard ordering the firing of a mustard-gas grenade launcher. It was, by all accounts, a chilling sound that sent Ann and Tim into a bit of a spin. Just what kind of Earth did they want to represent? They didn't want to airbrush out all our darker instincts, and yet at the same time, how real should this sound essay be?

The subject was debated that night at dinner with Linda, Carl and Murry Sidlin. While no solid conclusions were made then and there, the general consensus was that the sound essay should be a 'best of' rather than a 'worst of'. Soon afterwards, the policy was decided: a space-bound soundscape was not the place for apologetic soul-searching. With the best will in the world, how would it mean anything to an alien? Besides, it's not in our nature. How many of those irritating festive round-robin letters do you receive from your wider family? And how many of those letters note addiction, failures, adultery or debt? Hardly bloody any. Instead it's all promotions, prizes, deals and achievements.

No, humankind would have to smile for the camera. The Voyager Golden Record was going to show humanity on a good day.

CHAPTER FIVE

Now That's What I Call Music

'I mean, as an assignment, in one sense, it's laughably easy. You're making a selection of 90 minutes from all the music in the world. Clearly you could do that every year for your entire life and they'd all be great records.'

Tim Ferris

With Alan, Robert, Ann, Tim, Jon and Murry all sticking their oars in, there was a deluge of music on the table. Listening sessions continued. And although articles written about Voyager like to focus on disagreements, most of the participants remember these days with delight. Ann, speaking back in 2007 when the record was celebrating its 30th birthday, summed it up as 'thrilling, contentious months filled with beauty', saying: 'We took the responsibility to heart but luckily everyone involved had a sense of humour. There was some tension, yes, but also a lot of laughs. It was an honour and a deep source of pleasure.'

In an interview with a doctoral researcher named Stephanie Nelson in 1991 (for an article that would eventually be printed in *The Journal of Applied Communication Research* in 1993), Tim explained that there were three criteria at work when choosing the music: geographical diversity, economic diversity and 'good music'. Speaking to me in 2017, Tim said: 'Everyone worked on the music – all the principals, people who weren't principals, people who just heard about it. Lots of suggestions about the music came in and we didn't have any tsar to say this is a good idea or this isn't. We just kept listening to things and saying "yes" or "no" or "potentially", and pretty soon you have a collection that's five times more than you could fit on the record but is full of defensible stuff. You're starting to get somewhere.'

One particularly long debate took place on 14 May in a small office at the Smithsonian in Washington DC. 'When I

Need You' by Leo Sayer had just replaced the Eagles' 'Hotel California' at the top of the *Billboard* Hot 100. *Annie Hall* was doing well at the flicks. The world was about a week away from the premiere of the first *Star Wars* film. And the 7″ of 'God Save the Queen' by the Sex Pistols would go in sale in 13 days.

It was a Saturday. Wendy, Linda, Ann and Tim were all there, reviewing sounds at the Library of Congress. Towards the end of the day they met Carl, along with Murry Sidlin and his wife Debby. Inside the room was a hi-fi system and an enormous portrait of Louis Armstrong. They played some records. At times they listened in silence, at others they talked over one another. They enthused. They agreed. They disagreed. The meeting lasted until 3a.m.

A number of topics were debated that night. They discussed Native American music,[*] they discussed 'Summertime',[†] they discussed the blues and, with Satchmo eyeballing them all the time, they discussed Louis Armstrong. The Armstrong question, though, was never *whether* to send, it was simply *what* to send. Ann said: 'He was going to be there no matter what. You know it would have been over my dead body if we couldn't put Armstrong on there.'

The main issue on the table that night was the classical repertoire. Should they include more than one piece by the same composer at the expense of others? While Sidlin was 'vigorously opposed' to the idea, Carl felt that including two or more pieces composed by a single human being, would be a thought-provoking move. And as it was generally accepted that Beethoven and Bach represented the peak of the Western musical tradition, they seemed the logical choice. However, with such limited space, including more than one track by a single composer would effectively rule out including anything

---

[*] Carl shared with the group some input from University of Chicago expert Fred Eggens, who he had quizzed earlier in the day on the subject of Native American music.
[†] Jon Lomberg had already suggested using the Ella Fitzgerald–Louis Armstrong version.

by other celebrated greats such as Wagner or Debussy or Verdi or Rachmaninoff.* It wasn't a decision to be made lightly.

They moved on to other matters. At one point they played the Miles Davis recording of 'Summertime'. Carl felt this was the ideal choice as a representation of 'American' music on the Golden Record as it celebrated both black and white – written by a great American composer in George Gershwin and performed by a true pioneer in Miles Davis. Others in the room felt that black influence on popular music should be represented on its own, or 'without incumbent'.

Approaching something of an impasse, Sidlin put in a call to critic and Smithsonian jazz curator Martin Williams. In *Murmurs* Carl recalls this being a Sunday night, which may have been right – perhaps this 14 May meeting spilled over to the following night – but in any event, the call to Williams was at 11p.m. on either Saturday or Sunday night in mid-May 1977. Once Sidlin had confirmed that yes, he was indeed calling so late at night to ask what was the best jazz to send to ET, Williams offered some thoughts. 'Summertime', it seemed, was doomed.

'People sometimes talk about this project as if there were some sort of desperation to it,' says Tim. 'People fighting for their choices and everything. And you do get emotionally involved with music. But I think everyone recognised that if we don't use this terrific track we'll just use this other terrific track. I mean, as an assignment, in one sense, it's laughably easy. You're making a selection of 90 minutes from all the music in the world. Clearly you could do that every year for your entire life and they'd all be great records. What always worried me most was that if we don't get a spectacular result we're really going to look bad because the nature of the assignment is so easy.'

Ann's taste, which would exert a huge influence on NASA's mixtape, had been partially shaped by an older brother. Her parents had exposed her to folk music of the 1940s and '50s, and to Austrian singer Lotte Lenya.† But it was her brother

---

* By the sound of it, Rachmaninoff was never really in the running.
† Spike-kicking Rosa Klebb in *From Russia with Love* (1963).

who, from the age of six or seven, introduced her to rock 'n' roll.

'I loved all kinds of mostly popular music. I think that in the late '50s early '60s through the '70s, popular music was really distinguished and innovative – taking new territory. I loved Bob Dylan, I loved Motown, I loved soul music, I loved Otis Redding and I loved Marvin Gaye, I loved Martha and the Vandellas, Aretha Franklin, I mean I just loved all of it.'

As a teenager Ann had befriended Steven James, the nephew of Duke Ellington. She'd got to know the family, been hugged by Duke Ellington's sister Ruth, and she'd 'schlepped around' with the family to concerts of sacred music at local churches. She'd visited Max Roach and Abbey Lincoln's New York house for dinner, where Ann would sit under the piano, which just about everybody, it seemed to her, could play except her. Then, aged 19, she met Jonathan Cott from *Rolling Stone*.

'It was just an astonishing exposure to great music. He would take me to meet all these people and to listen intimately to their music in their homes and in all kinds of different settings.'

At this time, she was not a huge fan of classical music, being much more drawn to 'something with a backbeat and some soul'. However, by 1977 she had acquired a taste for other forms – for Stravinsky, for Beethoven – and she'd visited the Lincoln Center to listen to Pierre Boulez. Her taste was nothing if not eclectic, although she would admit that her knowledge of classical music was not comparable to her knowledge of blues, soul and folk music.

Back in March, when Carl had told Jon about the record, he'd made it clear that he wanted to satisfy the contemporary Earth-based audience by making it representative of the planet and species, not just of the Americans who were sending it. Jon, though, with his aliens-first hat on, wanted form and structure to be at the heart of the music.

'One of the first things that Carl and I had discussed years before was the notion that music might be as interstellar as physics,' he says. 'In other words, one of the things that people

like Frank had established as the axioms of the discipline was
that you couldn't communicate with somebody unless you
had something in common. And what would you have in
common? Well, you had the physical universe in common
and you had the laws of physics in common. So that's what
you use to establish communication. The series of prime
numbers. The Fibonacci series* – things that are not generated
naturally, but indicate intelligence.'

But, while pure mathematics could be the common ground
for establishing communication, it doesn't make for a very
interesting conversation. 'I mean, we both know prime
numbers. Great. But telling each other prime numbers back
and forth doesn't really get you very far.' Jon felt that a lot
of the arguments you could make for mathematics being
universal, you could make for *music* being universal. Carl had
shared that paper by von Hoerner with him by radio astronomer
Sebastian von Hoerner,† which argued that the notion of scale
and harmony were not arbitrary cultural constructs but that
they arose out of the physics of sound. Some frequencies
harmonise with each other, others damp each other out. 'So
that kind of amplitude and dissonance is not just something
we learn, it's really in the physics of the sound,' says Jon.

Another idea Jon floated was starting the record with
some explicit formal instruction in Earth's musical language.
Sound a note, play the same pitch an octave higher, introduce
fifths and thirds, a pentatonic scale, a major scale, a minor
scale, a chromatic scale. Then introduce more complex
scales, then follow this with a similar 'dictionary for rhythm
and timbre'.

However, Carl quickly axed this idea. They just didn't
have the space on the record to do it properly. They had to
devote as much available space to the music as possible. So
any thought of a rudimentary musical grammar lesson was
shelved.

---

* Every number after the first two is the sum of the two preceding
numbers.
† Another SETI mover-shaker.

Jon made more suggestions, on tape and by letter.* He chose some pieces to satisfy Carl's desire for a 'global' record, but his playlist is dominated by music with interesting structure or shape. In other words, Jon's priority was not that the music should be 'great', but that aliens could understand the music, and perhaps learn something interesting from it. Emotion was secondary to structure.

'Nobody likes to be sad but everybody likes sad songs,' he says. 'Music has a way of transmuting emotions to make them more bearable. On that level I don't know that we have much to say to the aliens because we don't know anything about emotions really – whether they even have them. So music that relied primarily on its emotional content seemed to me the music least likely for extraterrestrials to understand, whereas music that was more formal in nature would be a theme and variations, a rondeau, musical forms like verse and chorus, where there's an architecture that can be perceived. Those would be the pieces that at least the aliens could get some sense of "well, there's some structure here". So a Bach fugue will be unique … and a fugue is just a beautiful little puzzle in sound. All the information you need to know about that puzzle is there in the puzzle. In other words, there's nothing outside of the piece that you need to know in order to appreciate the piece. So that's kind of where the discussion started on music.'

Jon didn't worry too much about music from non-European traditions either, partly because he knew the team was already consulting with ethnomusicologists.

He writes: 'I proposed works from the Western classical tradition that I thought best met my criterion of accessibility-through-structure, including the gavotte en rondeau from J.S. Bach's Third Partita for Unaccompanied Violin, whose verse and chorus architecture had the clarity of a snowflake. Also Mozart's piano variations on the tune known in English

---

* In Appendix D of *Murmurs*, you can peruse the tracks from Jon's mixtape, designed for a one-hour record.

as "Twinkle twinkle little star", which seemed an appropriate tune. In both works a simple theme unfolding into a bouquet of variations demonstrates perfectly how a composer can take the simplest of melodies and weave it into webs of ever-increasing elegance and complexity.'

He suggested 'Ma Fin Est Mon Commencement' – the so-called Crab Canon by 14th-century composer Guillaume de Machaut – a musical palindrome for three voices, whose end is also its beginning. He also argued for Chaconne in D minor from Bach's Second Partita for Unaccompanied Violin, a piece of 'limitless depth and profundity' that was never really in the running being some 14 minutes long.

Sidlin, Carl's conductor friend, had already suggested the Prelude and Fugue from Bach's *Well-Tempered Clavier*, and Jon put forward the performance by Canadian pianist Glenn Gould, who he sometimes ran into at the Xerox machine in the CBC radio building.* His tape also included 'Summertime' (the Ella Fitzgerald–Louis Armstrong recording) and 'Sgt Pepper's Lonely Hearts Club Band (reprise)' by The Beatles.

'Had it been left up to me, I might have chucked all the rest of the Beethoven and other Bach pieces, and let the Chaconne and Mozart aria stand for the Western tradition of classical music,' Jon wrote. 'Nothing could represent us better.'

Tim, however, was most concerned with making a good record. That was his overriding ambition. I asked him whether, back in 1977, he had ever felt overwhelmed by a sense of responsibility of selecting music to represent Earth. He replied: 'Well, there are lots of fields where I would not have known enough, or had sufficient maturity, to be responsible in accepting such an assignment. Economics, for instance. I wouldn't have known enough about economics,

---

* They were both freelancers mixing radio programmes with engineer Lorne Tulk, and they had to type and Xerox their own scripts. Apparently Glenn operated the copy machine without removing his gloves.

or many areas of history or social policy. I was just too ignorant of those things. But science and music were not in that category. I was working as a writer … I had an apartment that had a home office stuck out in the middle of these treetops – it was a really good place to work. And then we would have music listening sessions at either my place in New York or at Carl's house in Ithaca. It was fun – you're just listening to music and then you're gathering comments.'

He chuckled to himself and continued: 'A lot of the selections we were listening to, the time wasn't listed on the medium and I was supposed to time all these tracks and I kept forgetting to time them. That was my constant flaw. Carl's was that he kept forgetting that time is a six-based system not 10-based, so he kept dividing seconds by 10. So we each had our blind spots. We knew each other pretty well and we'd done a lot of work by then. We'd edited a lot of each other's copy and everything – we worked pretty smoothly together.'

Ann and Tim paid the first of several visits to Alan Lomax's place. He lived in an apartment in uptown New York, 215 West 98th Street, 'a warren of 100,000 LP records'. One of the first things he put on the turntable when they arrived was 'Melancholy Blues' by Louis Armstrong.

Armstrong had passed away in Queens six summers before at the age of 69. Born in New Orleans, his career had started on street corners, singing and playing the cornet. He'd been sent to the Colored Waifs' Home by the New Orleans Juvenile Court for firing his stepfather's handgun back on New Year's Eve in 1912. Already a self-taught musician, it was here that he developed his skills, playing with the home's own band and eventually becoming band leader. By 1925 he was living and performing in Chicago, and his reputation was beginning to spread beyond the jazz community. He had starred in leading bands, performed at big venues, and played for both black and white audiences. He was known for his emotional, expressive playing, effervescent jive delivered in distinctive gravelly tones, and in contemporary PR material was being billed as the

'World's Greatest'. It was around then that he made his first recordings for the Okeh jazz label under his own name, with his 'Hot Five' and 'Hot Seven' groups.*

Jazz historians generally agree that this period in Armstrong's career changed the direction of popular jazz music. The focus was torn away from the traditional New Orleans collective improv, to numbers dominated by individual solos. And it wasn't just soloing; Armstrong was trying out new rhythms and arrangements, and his singing on records such as 'Heebie Jeebies' would popularise scat. These records shook up jazz, influencing his contemporaries and generations to come.

In among 12 songs with his 'Hot Seven' was a number called 'Melancholy Blues', recorded in Chicago on 11 May 1927. The fact that it was chosen above all others from this period is all down to Alan Lomax and his warren of LPs.

Alan played them one song after another. After 'Melancholy Blues' came 'Dark Was the Night' by Blind Willie Johnson, then a Georgian men's chorus. Throughout he enthused about his theories of cantometrics, about how the Voyager record could be used as a platform to show the development of human culture through music. One highlight that would also make the final cut was a haunting Bulgarian folk song called 'Izlel ye Delyu Haydutin'. It tells the story of a folk hero who harasses and badgers occupying troops, a tale of resistance to an outside invader. It was performed by Valya

---

* According to Brian Harker's *Louis Armstrong's Hot Five and Hot Seven Recordings* (2011, Oxford University Press), the Hot Five band included Kid Ory (trombone), Johnny Dodds (clarinet), Johnny St Cyr (banjo), Louis' wife Lil on piano, and usually no drummer. Over a 12-month period starting in November 1925, this quintet produced 24 records. The Hot Seven sessions took place in May 1927, producing a further 12 songs. The Seven line-up included the Hot Five lot (although John Thomas was now on trombone as Kid Ory was off touring), plus Baby Dodds (Johnny's brother) on drums and Pete Briggs on tuba.

Balkanska,[*] backed by gaida[†] players Lazar Kanevski and Stephan Zahmanov, and recorded by Martin Koenig and Ethel Raim in Smolyan, Bulgaria in 1968. When Alan first played it for Ann, she was moved to dance. Lomax leaned forwards, grinned, called her 'honey', then explained that this was the sound of agricultural communities, the first people who had enough to eat.

Tim says: 'Alan was a good producer. His dad was a good producer. He was very valuable to me on the record … We left on pretty good terms but he always felt that he should have had more to say about the choices rather than just coming up with material.'

Ann says: '[Alan] was a very cranky, cantankerous, difficult man. But he was a genius of ethnomusicology. Some of the stuff that he gave us turned out to be other than what he told us it was. You know, he told us it was one thing, and then we find out later on that it was something else, but he was hugely important to the music.'

This uneasy interface between Lomax and the record team is also where some errors crept into Voyager's liner notes, many of which would last for decades. They were first printed in *Murmurs*, and were continually reprinted and shared by NASA online and in print. It wasn't until the crowdfunded 40th-anniversary project by Ozma Records, which succeeded in bringing out a deluxe reissue of the Golden Record in 2017, that some errors and missing pieces of information were uncovered and put right.

David Pescovitz – who co-produced the Ozma Records project with San Francisco record-store manager Tim Daly and music-packaging designer Lawrence Azerrad – explained that in some cases Alan had provided the Voyager team with unreleased recordings on tape reels that just simply didn't carry any information. He found this out when viewing scans of the

---

[*] Valya was born in 1942 in a hamlet near the village of Arda, Smolyan Province. She is a legend in her home country for her wide repertoire of Balkan folksongs.

[†] A bagpipe-like instrument.

original reels and tape boxes from Lomax's collection that are now stored in the US Library of Congress. In the case of the Solomon Islands track, this had been known simply as 'Solomon Islands Panpipes'. David called the Solomon Islands Broadcasting Corporation (SIBC), the local radio station that had made the original recording, to see what he could find out. The librarian knew about the record, sure, but didn't know who the performers were. David's next step was to seek the help of anthropologists and musicologists in the region, including Martin Hadlow, a communications professor at the University of Queensland who had done extensive research at the SIBC. None of the leads paid off until months later when Hadlow happened to attend a meeting at the SIBC and mentioned the Voyager record. A young woman working there overheard and revealed that she was related to the original musicians. Several weeks later, Pescovitz received a recounting of the song's recording as told by Isaac Smith Houmaawa'i, leader of the group. So a song, formerly known simply as Solomon Islands Panpipes, was now correctly labelled 'Naranaratana Kookokoo' ('The Cry of the Megapode Bird') as performed by Maniasinimae and Taumaetarau Chieftain Tribe of Oloha and Palasu'u Village Community in Small Malaita.

Another piece that came straight from Alan Lomax was 'Senegalese Percussion', an atonal piece that's all about rhythm – the instruments are drums, bells and flutes, but even the flute-like instruments really serve as another drum. If you go online to read some official NASA history of the Voyager record, it will usually still identify one of the tracks as 'Senegalese Percussion', or sometimes by the name 'Tchenhoukoumen'. Both are wrong. It's actually from Benin and called 'Cengunmé'. The errors this time were revealed to David and his team by Charles Duvelle,[*] the musicologist

---

[*] Charles Duvelle, who died in 2017, was born in Paris but spent most of his childhood in Laos, Vietnam and Cambodia. He studied music at the Conservatoire de Paris, then in 1959 was hired to work with and record African radio programmes. He founded the Ocora label in 1962, which grew into one of the most extensive catalogues of recorded traditional music.

who had collected the recording in the field. The title of the song was incorrectly listed on the jacket of the LP that Lomax loaned to Tim Ferris – from which they sourced the song. David said: 'I know which LP that was because I have a receipt that Lomax made Tim sign when he loaned him the records.'

'Cengunmé' was performed by Mahi musicians and recorded by Duvelle in January 1963 in Savalou, Benin, West Africa – about 1,000 miles from Senegal.

Tim told me: 'I read somewhere that Alan had contributed three-quarters of the record or something, and that's nonsense. But he did contribute a lot. He was a unique individual … He was not the best person in the world at working with others. He had some frustrations in life, he was broke all the time. And we tried to help him out a bit, but he always wanted a bigger role and more recognition.'

As we explore more of the songs in the running for Voyager 1 and 2, it's worth introducing Folkways Records. To give you a flavour of the incredibly varied back catalogue at Folkways, here are a handful of titles picked at random: *Sweet Thunder: Black Poetry* by Nancy Dupree (FW09787, 1977); *Electronic Music* by the University of Toronto Electronic Music Studio (FW33436, 1967); *Koto: Music of the One-string Ichigenkin* by Isshi Yamada (FW08746, 1967); and *Jewish Folk Songs* by Ruth Rubin (FW08740, 1959). The label was founded by Polish-American recording engineer Moses Asch in 1948, with the aim of documenting the entire world of sound. The 2,168 titles Asch released on Folkways* included traditional and contemporary music, spoken word in many languages, and documentary recordings of individuals, communities and events.

One Folkways track that was in the process of graduating from the Voyager team's longlist to the shortlist was Native American 'Navajo Night Chant'. 'Night Chant, Yeibichai Dance', to give it its full title, was originally issued in 1951 on

---

* The entire Folkways catalogue was acquired by the Smithsonian Institution in 1987, which today keeps every recording in print.

the Ethnic Folkways Library record *Music of the American Indians of the Southwest* (FW04420), with original liner notes by Willard Rhodes, the then-associate professor of music at Columbia University. The track was recorded by Rhodes in Pine Springs, Arizona, in the summer of 1942. The night chant was one of 35 major Navajo ceremonies, part of a rite-of-passage ceremony initiating boys and girls into the tribe's ceremonies, accompanied by gourd rattles. At the time that the Voyager record left Earth, it was widely known that the chant was recorded by Willard Rhodes. Now, thanks to the researchers at Ozma Records, the tracklist can include the names of the performers: Ambrose Roan Horse, Chester Roan and Tom Roan.

The Voyager records would also have two pieces representing parts of the former Soviet Union, both from the Caucasus. There was a Georgian chorus championed by Lomax, and a piece often labelled 'Azerbaijan Bagpipes' or 'Ugam'. It is actually called 'Muğam' and it doesn't have any bagpipes. It was performed by folk musician Kamil Jalilov, playing the recorder-like balaban, a cylindrical-bore double-reed wind instrument, and was recorded by Radio Moscow circa 1950. The errors came from the original liner notes of Folkways record *Folk Music of the U.S.S.R.** (FW04535) in 1960.

***

---

* The LP was compiled by the American composer, world-music expert and theorist Henry Cowell (1897-1965). Cowell would write about and experiment with harmonic rhythm – the rate chords change in relation to the rate of notes. Indeed in 1930 he commissioned Léon Theremin to invent the Rhythmicon – the first proto drum machine. Henry was raised by his mother, Clarissa Dixon, author of the early feminist novel *Janet and Her Dear Phebe*. Later in his career, and following a four-year stretch in San Quentin for illegal sexual acts, Cowell mentored many famous composers (including Burt Bacharach), also writing and compiling for Folkways Records.

After their week in Washington DC, Ann and Tim had returned to New York with around 50 sounds to play with. These weren't going to be plopped together any old how; they would need to be crafted, edited and mixed to create a coherent sound essay further down the track. And in the strict narrative sense of this story, the actual mixing wouldn't begin for a while yet. But to explore the sound essay further, I'm going to take you through its early movements.

The sound essay goes by the name of the 'Sounds of Earth'. Press play on Voyager's 'Sounds of Earth' and it lurches into the 'pre-life' phase. A series of rumblings – volcanoes, earthquakes and thunder – are followed by gurgling mud pots, then sounds of wind, rain and surf. Next comes life, with crickets and frogs, the majority of sounds taken from the CBS library.* The essay moves up the food chain – from insects to birds, to hyenas, elephants, chimpanzees and dogs – before, still only a few minutes in, we hear our first human sounds: footsteps, mixed with a heartbeat.

Then the first human steps up to the interstellar mic. And what does he do? He laughs. That's right, the first sound generated by human vocal chords[†] within the Voyager record sound essay is not speech, not a shout nor a murmur, but laughter. This all seems pleasingly appropriate, what with laughter being more or less unique to us humans. Yes, experiments have shown that if you tickle baby rats you can hear what could be described as a kind of ultrasonic laughter. But still, if we were in a classroom with every other species on Earth, each showing off some unique attribute during an eternal show-and-tell, we might well choose laughter to show the class, or possibly opposable thumbs.

---

* Apart from one adult male cricket, who was recorded showing off to a female by Dr Ronald R. Hoy at the Langmuir Laboratory at Cornell University.

† However, the sound essay section is preceded on the record by spoken-word greetings so, in fact, an alien listener would have heard human speech prior to the laughter in the sound essay.

The laugh in question is rather odd. It's a short, high-pitched staccato sound of someone obviously having hysterics, followed by one of those post-mirth recovery 'wahoos'. It isn't a perfect, archetypal laugh. It's quite a strange, atypical laugh. There's nothing wrong with it, you understand, it's just not that representative. It's a laugh, no doubt about it. It sounds free, easy and genuine, but you wouldn't necessarily pick it out of a laugh line-up. And it's quite easy to miss in some of the lower-quality online versions of the record. You can tell it's a man doing the laughing, and that's about it.

In *Murmurs* they don't mention where the laugh came from. It's simply grouped together in the soundscape playlist as 'Footsteps, Heartbeats and Laughter'. Then in 2017, during the build-up to Voyager's 40th-birthday shenanigans, Adrienne LaFrance wrote a piece about the laugh for the *Atlantic*. She wanted to know whose it was, and why it seemed to be missing from NASA's official Soundcloud version of the record. Turns out it wasn't missing at all – it was just very hard to hear because of the quality of that particular source. Nevertheless, Adrienne still wanted to know who was doing the laughing and, despite lots of digging, couldn't find anybody who could answer that question. NASA didn't know, the Library of Congress couldn't help, the Carl Sagan Institute at Cornell was nonplussed, and Jet Propulsion Laboratory-Caltech had no knowledge.

Then JPL's Senior Storyteller, science writer Elizabeth Landau, got back in touch to say she'd met with Ann Druyan earlier that day who told her the laugh belonged to Carl Sagan. Ah, good. That all seems to fit. But no, wait. Tim Ferris was also contacted about it. He had no memory of it being Carl, felt that it didn't sound like Carl, and that it was too good a fit for the story to be true. Oh Lord.

I can't say with any certainty whether it is Carl or not. I've listened to it a lot. I've listened to it over and over, then compared it with Carl laughing on YouTube clips and the like. I think it could be him, but it could easily not be him too. The team did do some of the sound recordings themselves – on the fly and in the field during those hectic

weeks of May 1977 – so it's perfectly plausible that it *is* Carl
doing the laughing. Plus, if it is Carl, that would explain why
this slightly odd, atypical laugh made the final mix,* rather
than a better example from a sound library's vaults. And
certainly, in a blog post dating from 2007, long before
Adrienne's piece, Ann writes about it being Carl's laugh. We
can also look to the original reel-to-reel tapes that survive
within Sony's archive for further evidence. These come from
towards the end of the record's mixing and mastering process.
One of the tape boxes is scrawled with numbered tracks, in
various different hands. The first half of the list comprises
sound effects provided by a sound library. The second half is
all sounds that we know were created by the team – and
sandwiched within this list is the word 'laughter'. Still,
whoever it is, don't you think it's rather splendid that in this
ultra-serious, rather worthy endeavour to transmit humanity's
soul to the cosmos, we start off with a guffaw?

We'd met insects, said hello to birds and listened to some
chimps making a hullabaloo. Next comes the desolate howl
of a wild dog, heard against a bleak rasping wind, all sounding
very Hammer horror, before giving way to humans (by now
about six minutes in) whose aforementioned hysterical
laughter lightens the mood. Just after the laughter, with the
heartbeat playing behind in the mix, comes 'Fire and Speech'.
We hear the crackling of flames, and then the first talking.
Specifically this is the distinctive click language of the !Kung
Kalahari bushmen, although the words are not spoken by the
hunter-gatherers themselves but by noted anthropologist and
Toronto professor Richard Lee, who studied the !Kung for
many years and authored the influential work *Man the Hunter*.[†]

---

* You can read Adrienne's piece via theatlantic.com. It was published
in mid-2017.

[†] A slightly unfair title as both men and women tend to share the
hunting and gathering duties. Plus gathering (mainly mongongo
nuts) provided a higher percentage of the diet than the macho
hunting. *Woman the Gatherer* would have been better. Incidentally,
Sagan would write about the !Kung way of life in 1995's *The
Demon-Haunted World*.

Next the team wanted to illustrate that key stopping-point on human evolution – the use of stone tools. They wanted the sound of tools being fashioned, but nothing was coming up in the sound-library vaults. Carl went wandering the streets of Midtown New York to find suitable rocks to bang together. I don't know if you've ever hunted for flint tools, but the streets of New York is not where I would start. He couldn't find anything suitable, so Carl consulted the Rolodex again and called Alexander Marshack (from the Museum of Comparative Zoology at Harvard) to get the low-down on how tools were made. Linda got hold of some flint samples from Ralph Solecki[*] at Columbia University – indeed, Ralph went further, also providing gloves and goggles[†] – and they set to work, recording the sound of stones hitting stones.

Part 12 of the sound essay is known as 'Tame Dog'. Unlike the wild dog from a few minutes before, this dog sounds like an ordinary domesticated animal and was deliberately placed in the chronology after the first sounds of human activity in order to send this message to ET: one wild dog, plus humans with tools, equals one tame dog.[‡]

'Tame Dog' is followed by more agricultural sounds: herding sheep, a working blacksmith's, sawing and a tractor. Ann writes in *Murmurs* how they had toyed with sounds of roosters and cows too, but these were abandoned as sounding too stagey. With the human audience in mind, that was probably a smart decision. There might have been barely stifled guffaws had they gone full 'Old Macdonald'.

After the sounds of agriculture and first machinery, come the first sounds of long-distance communication. The team

---

[*] An American archaeologist, best known for his excavations at the Neanderthal site at Shanidar Cave in Iraq.

[†] If you have ever taken part in any experimental archaeology – where archaeologists attempt to recreate some archaic working method in order to find out how an object was produced – then you'll know that flint knapping is no walk in the park.

[‡] Dogs also turn up in the picture sequence. As do the !Kung.

had already decided they wanted to include Morse, but what message? Carl immediately suggested 'Ad astra per aspera' (to the stars through difficulties), and the message was tapped out by Robert R. Schoppe, a radio operator at CBS.

Next comes the 'Transportation Sequence', where we hear ships, horses and carts, trains, trucks, tractors, buses, cars, an F-111 fly-by and the sound of a Saturn V rocket launching Apollo 17 in December 1972.

★★★

Alongside the music team, the picture team and the sound-essay sub-team, sat the language or greetings sub-team. The brief for this group was to record an array of greetings in as many human languages as possible. This was to be a literal 'Hello, universe'. In some ways this is the most outward-facing aspect of the record, in other ways the most pointless. It would also cause a hell of a lot of trouble.

The idea to include any greetings was open to question in the first place. Everyone knew that even the most gifted alien linguist would be unable to decipher the sounds – there were never any plans to send any kind of audio Rosetta Stone. Nevertheless, it seemed a perfectly pleasant idea to have at least some form of spoken 'hello' on the record, and the moment that decision was made, it seemed too open to criticism merely to say 'hello' in the predominant Western language of American English. So they decided to say 'hello' in multiple languages, and soon the ambition was to cram on as many languages as they could. This was to be a global 'yo'. And the moment *that* idea was on the table, it wasn't much of a stretch to argue that a smattering of some non-human language should be included too.

Ann recalls: 'When [Carl] was talking about the greetings and language, he was saying we want to include all these greetings in many different languages … And I looked at him and said: "Only human languages?" And he gave me this look which was one of such affirmation – that I was the right woman for the job.'

So around the same time Ann was trawling sound libraries for hyenas, she called Dr Roger Payne at Rockefeller University. Roger and his wife Katie were among the first people to study the communications of humpback whales, with the hope of interpreting them. Ann explained all about their interstellar global yo, and their wish to include whale greetings alongside the human ones. Roger said: 'Where have you been all my life?' He could barely contain his enthusiasm, promising to bring everything he had, all the recordings he had made during years of study in the field. Indeed, he had a favourite one already in mind, a recording made off the coast of Bermuda in 1970.

The greetings/languages team started out with pretty lofty ambitions. But in the end, with time running out, the majority were compiled in a relatively last-minute scramble by Cornell staffers and PR bods. The languages of Earth, humanity's great global greeting to the cosmos, were brought together in the main by dipping into the cultures represented in Cornell's student and faculty body that summer. Why? Because politicians don't know when to shut up.

If you look at an online Golden Record playlist right now, the majority of spoken-word greetings are divided into two chapters. The first in the production timeline is known as 'UN Greetings', and was Sagan's bright idea. He was in New York. Naturally enough, he thought to himself: 'Where can we quickly find lots of people who speak different languages in New York? Why the United Nations, of course!' Sagan pictured a kind of open-mic operation, where delegates could be invited to simply say 'hello' in their own language. Excellent plan. What could possibly go wrong?

In *Murmurs*, Sagan writes about how he first asked the US delegation for help. It sounded simple enough, but the magnitude of this one-time-only chance to greet the cosmos was too great for them. They didn't have the right to take that responsibility. A global greeting to be preserved *forever*? This was too far above and beyond their pay grade. Plus, it was weird. So Carl want to the UN's Outer Space Committee, as he knew some of them personally. Their feeling was: 'Yes we can do this! … But we can't initiate it. We need to be acting on orders from above.' So Carl went back to the US delegation again. They made it clear: 'We'll do it, sure, but we need to be *told* to do it by the State Department.'

Already pulling his hair out at this chicken-and-egg conundrum, Carl was then told off. He should *not* have gone to the Outer Space Committee directly. He had said too much. Now everyone knew that this rather strange interstellar monument was an American-funded project, and many delegates would refuse to co-operate or might even attempt to block the whole thing from happening for that reason alone. And what was with this free-and-easy, counterculture-esque idea for an open recording session anyway? That wouldn't work. Not every UN diplomat was in town every day. What happened to all those people who happened not to be in on the day the recording took place? Pure chance would omit them, their culture and language from the record. Think of the political ramifications. Eventually, the only useful suggestion on the table was that Carl could record each one of the UN's Outer Space Committee saying 'hello'. But, as noted in William Poundstone's Sagan biography, this would fail to achieve what they set out to do; while many different languages were represented by the members of the committee, there were sizeable gaps. There was no Chinese member, for example, and the committee was nearly all male. Then the committee revealed that they would first have to vote about the possibility of saying 'hello', and the next meeting was scheduled for late June, which was too late.

If we equate Carl's attempts to record greetings spoken by members of the UN to Carl buying a car, the situation was currently playing out like this:

Carl enters car dealership.

Carl: 'I'd like something with four doors and plenty of space. It needs to be a family car.'

UN Outer Space Committee: 'We have a two-seater convertible.'

Carl: 'Hmm. It's not what I was after, but I'm desperate and short of time. It will have to do.'

UN Outer Space Committee: 'Great! I'll get the paperwork. By the way were you hoping to drive it? It's just this car doesn't start.'

At one point Carl was even asked if the launch could be postponed.

★★★

Once Jon had found a space to work at Cornell, one of the first things he did was collate all the suggested topics into a master list. He wrote out these subjects on a large piece of white paper and taped it to the wall. Then he started looking through books and magazines. He writes: 'I tried to open myself completely to the flood of images that passed across my desk, looking for those images that promoted themselves as candidates.'

Jon and Wendy went on trips around Cornell and to local public libraries, plundering coffee-table and picture books like *Birds of North America*, the influential *Family of Man* and *The Age of Steam*. One of the Cornell staffers brought in about 20 years' worth of *National Geographic* magazines back to 1958 that had been sitting in her garage, 'unaware that their fate was to be perused and cut to pieces for our candidate file'.

They had already decided to avoid politics and religion. Another important early decision was not to include artwork. Again, put yourself in their shoes and it seems a logical decision. They already had enough on their plate, choosing diagrams and photographs that were easy for aliens to interpret, without muddying the water with art. The casting net was quite wide enough, thank you very much. Plus, as Jon reasoned, human art was already being well represented over in the music section. The playlist was art. The pictures would be the factual bit.

This 'no art' policy would also inform the choices of the photographs themselves. When you look through the Voyager

Golden Record photographs today, the ones that made the final cut, some are undoubtedly beautiful – Ansel Adams's 'The Tetons and the Snake River', for example* – but many seem prosaic or utilitarian. This is precisely because the overriding factor was the information the image could convey, rather than any aesthetic value.

Back in the Cornell work room, if Jon, Wendy or any of the others found an image that fitted one of the required topics, Jon noted down the location on his wall chart. 'After a few days some topics had many candidates, others only one, some none at all … It was clear that the *National Geographic Magazine* represented a huge resource for us,' he writes. 'They do, month after month, the same thing we were attempting to do in creating a portrait of the lands and inhabitants of our planet.'

Jon contacted the NGS office and, with the help of a picture editor named Jon Schneeberger, received permission to use the images as well as access to transparencies and even unpublished images from the magazine's archives.†

'As the images began to pile up, they began to sort themselves into groups. It was like creating a jigsaw puzzle, with a planet full of possible pieces. The [breadth of] considerations that guided the selection was a hall of mirrors, with a bewildering array of criteria for selection – technical, cognitive, political, aesthetic, financial, legal, practical … We mostly flew by the seat of our pants, selecting and rejecting candidate images case by case, letting the sequence grow itself.'

Jon shared with me a photograph of his original wall chart. On it were noted a number of categories, with references to the images found so far that fitted the bill. So next to '12. Coast', John had written 'Hawaii Guide, Cover; *Drama of the Oceans* p.195; *The Beautiful Land* p.17; *National Geographic* Jan 1968, p.102–3; *NGS* October '66, p.518–19; UN – Greek Boat'.

---

* Photographed at Grand Teton National Park, Wyoming in 1942.
† Of the 120 images in the Voyager sequence, 28 came from NGS.

They also contacted Tom Prendergast, the Picture Librarian at UN Headquarters in New York. 'Tom spent two days helping me look through the photo library,' Jon writes, 'looking through their slide collection of images of people from around the world, trying to decide which citizens of Earth would take the big ride.'

This approach to the project was shaped in part by a letter that sci-fi author Robert A. Heinlein had sent to Carl back in December 1976,[*] which Carl had in turn shared with the team. It's a long, thought-provoking letter, full of interesting ideas. Heinlein sets out his stall at the start. His intentions are – instead of offering the sort of ideas that Carl will already be receiving from other quarters – to offer unsound advice, illogical ideas, as wild as he could dream up, to play devil's advocate. He wanted to test the walls of the room in which they were working, to put a stethoscope to their assumptions. He talks about senses that aliens might have that we don't. He talks about precognition, he ruminates over the difficulties of communicating with some other intelligent being when there is no common cultural background.[†] He critiques previous attempts at extraterrestrial communication, by radio, by geometrical diagrams, by 'carefully built-up codes that establish a number system before branching into concepts assumed to be universal truths, such as atomic weights of elements'. He is plainly thinking at least in part of the Arecibo message, and he discloses that he – a human with shared cultural background, education and experience as a member of a Naval Code and Cipher Board – had never successfully managed to break these codes. 'I found that I could always read them after I read the English explanation. Never before.' His conclusion? That this message is not to extraterrestrials at all, but a time capsule for our descendants.

---

[*] At the time he might well have been working on his gloriously pulpy romp *The Number of the Beast*, which came out in 1980.
[†] A theme he explored in his 1961 novel *Stranger in a Strange Land* about a born-and-raised Martian who comes to Earth.

Heinlein also questions the assumption that ET can see. A life form that evolved under a different sun might not be able to see in the same way we understand it. Human eyes, he writes, cannot see the 'ruled grooves' in a spectroscopic grating or the heat waves in a pitch-dark room, because we evolved here, beneath our sun. He suggests, therefore, that any message on a plaque should be 'etched or grooved' to be seen over as wide a range of the spectrum as possible, to cater for alien eyesight that operates outside the remit of what we call 'visible light'.

It's a playful and provocative letter. I also note with interest two uses of the word 'groove' in fairly quick succession. And I wonder too, if Carl shared this document with Frank, whether it might have sent 'grooves–vinyl–record' word-association sparks firing through Frank's brain. What's without doubt is that the letter arrived at the Space Sciences Building at Cornell a few days before Christmas 1976, and Frank suggested his heavy-metal record in Honolulu in late January 1977. But whether Heinlein can claim any credit for the initial idea or not, his letter would certainly be enormously influential in guiding Jon's thinking on the kinds of pictures they should be selecting for the record.

'I was especially concerned by the idea that the concept of picture as we understand it may be mysterious to an alien,' Jon writes. 'We wanted the pictures to be understandable but we also wanted them to contain as much information as possible. Those two demands might be mutually exclusive.'

Heinlein's letter prompted Jon to focus on images where form follows function, where the look of the object should help the viewer understand what it is. A bicycle, for example, would be easier to understand than a car since the mechanics are all exposed. A radio telescope is shaped to catch and focus long wavelength radio waves, so its design is guided solely by that function.

At one point Jon also sought the advice of Charles and Ray Eames. The Eames were famous not only for their industrial and furniture design, but for their whole approach to design. Maybe they could offer some insight into designing an interstellar message, Jon thought. He had watched their scientific film *Powers of 10* over and over again when he had worked at the Ontario Science Center, and in 1975 he had built a large three-dimensional model galaxy for their astronomy display.

'Every few hours I would take a break and wander to the kiosk where *Powers of 10* was being shown on an endless loop,' Jon writes. 'The film started with a man sleeping on a blanket. Each successive image was 10 times larger than the previous one, pulling back until the Earth, solar system, galaxy and known universe filled the frame. Then it dived back down to the sleeping man and continued in decreasing powers of 10, ending at the nucleus of an atom. This film itself inspired exactly the same kind of cosmic perspective that Carl so often urged. It seemed to have the germ of an idea for how material on the Voyager Record might be organized.'

Jon called Charles. 'I outlined our project and said I would very much appreciate his suggestions at how our image sequence might be envisioned.' Charles was silent for just a moment, then he became really rather cross. How *dare* they do something like this. They should *not* be doing this. This should take *years*. This should involve a major international design committee – not just be thrown together in a last-minute rush!

'Maybe,' Jon countered, 'but we have almost no money and no time.'

Jon assured Charles they would do the best they could. Would he please help? Eames grew angrier. No, he would not help.

#Click#

During the course of the project Jon would make lots of calls. Some people didn't believe him when he told them what they were up to. Others thought it was just a 'silly waste of time'. One photographer's agent said 'this is the screwiest project I ever heard of,' before genially granting him permission. 'But Eames was the only one who got mad at us for even making the attempt,' writes Jon.

★★★

By now – between Jon and Wendy, Frank, Valentin Boriakoff, George Helou and Amahl Shakhashiri of NAIC – they more or less filled a room in the Space Sciences Building with books and magazines. The array of titles that would give up their wares to Voyager is one of the pleasures of the record's story. If all the images had come from a single picture library, it might not have been so much fun. The fact that Voyager also serves as a snapshot of mid-century, mass-market and educational coffee-table publishing is a fact to be celebrated. There's a picture of man using a drill (#96), for example, which comes from *Gem Cutting* by John Sinkankas,* chosen as it shows in close-up a human figure using its hands to manipulate smaller objects. There's a picture of Oxford (#91) from *C.S. Lewis: Images of His World* by Douglas R. Gilbert. There's a picture of someone cooking fish, and a picture of a Chinese dinner party, both of which were found in two titles in the Time-Life 'Foods of the World' series. The picture of an eagle (#58) comes from the 1974 work *Doñana: Spain's Wildlife Wilderness*. The photographs showing fertilised ova were taken from the 1965 work *A Child is Born* by Lennart Nilsson. The human-anatomy diagrams were adapted from *The World Book Encyclopedia*. The diagram of human reproductive organs comes from *Life: Cells, Organisms, Populations* by E.O. Wilson.

As the team chose the photographs, they would slot them into a broadly evolutionary sequence. They wanted diverse landscapes, plants and animals, humans engaged in a wide range of activities, family and community, structures, agriculture, and the planet Earth itself as seen from space. At this point Jon still hoped the sound essay and pictures could work hand in glove, each aiding the interpretation of the other, although this was looking increasingly unlikely. So instead, he focused on how best to arrange the photographs.

'We wanted to assemble them so there was an overarching story,' he says. 'And then within that story there are smaller

---

* Frank is a keen gem cutter and this obscure volume came from his own library.

stories, so that the individual components would make some kind of sense so you got the flow even if there was a piece or two that you couldn't understand … So this is a sequence showing vehicles, for example. Well, if there's one vehicle you couldn't recognise – "you" being the extraterrestrial – if it was in a sequence of *other* vehicles, that would help you understand "oh, this must be some kind of vehicle". Whereas if you just threw the pictures in a random assortment – if living things were mixed in with machines were mixed in with minerals – it might be harder for an alien reader to understand what was what.'

However, as the hundreds of possible images piled up, a particularly knotty problem was now facing the picture team: scale. They felt they needed to give everything a scale. They wanted some of the photographs to include measurements in distance, in weight and even in time. But how would you do that? It would be no good adding a line and writing '10 centimetres' next to it – that would be meaningless to an alien. They wouldn't recognise what the symbols meant or what 'cm' stood for. How do you give a workable scale to an alien? How do you fix a scale of time, distance and weight in two dimensions, without any common language?

'Frank? *Frank!* Get your butt in here!'

CHAPTER SIX

The Hydrogen Key

'The movements of the heavens are nothing except a certain ever-
lasting polyphony – intelligible, not audible...'

Johannes Kepler

The first time Laurie Spiegel saw a synth was in 1969. She was in Morton Subotnick's studio above the Bleeker Street Cinema, Manhattan. It was love at first sight. It was an original Buchla 100, the very instrument Mort had used to record *Silver Apples of the Moon*[*] two years before. It was an epiphany for Spiegel. Mort moved west, but the Buchla stayed in the city, ending up at a basement studio at the NYU Film School. Laurie continued to use it, but gradually grew frustrated with its limitations. Then in 1973 she began working at Bell Telephone Laboratories, giving her access to computers that allowed much more 'complex, reproducible and stable control of sound'.

The only original electronic 'music' aboard Voyager is a piece called 'Music of the Spheres', positioned at the start of the 'Sounds of Earth' sound essay. It is essentially a fabric made of ever-changing computer-generated tones, each one representing a planet's orbit rising and falling as it circumnavigates the Sun. It was inspired by Johannes Kepler's 17th-century tract *Harmonices Mundi*[†] (*The Harmony of the*

---

[*] One of the first commercially successful electronic music LPs.
[†] He wrote: 'The movements of the heavens are nothing except a certain ever-lasting polyphony (intelligible, not audible) ... Hence it should no longer seem strange that man, the image of his Creator, has finally discovered the art of singing polyphonically, which was unknown to the ancients. With this symphony of voices man can play through the eternity of time in less than an hour, and can taste in small measure the delight of God the Supreme Artist...'

World), in which, alongside a third law of planetary motion,[*]
he imagined the movements of the planets as a series of voices,
singing polyphonically.

In late 1976, and completely independent from what was
going on with Carl and the Voyager team, Yale professors
Willie Ruff and John Rogers had approached Laurie with the
idea of using a computer to realise Kepler's vision. Laurie was
the perfect person at the perfect time. In the mid-1970s she
had been discussing with her Bell Labs superior Max Mathews
the idea that, just as visual display of scientific data using
then-brand-new computer graphics was shedding new light
on that data, perhaps it would be possible to gain different
insights from an audio display of scientific data.

'So when Willie had suggested realising Kepler's idea, it
was already right up my alley and it completely made sense
for me to do it … His idea of astronomical phenomena being
made audible was a natural for me to want to try, and the
computer made it possible, providing a bridge between
physical reality and audio, with math as the translating
Rosetta Stone.'

Willie gave Laurie some astronomical data on the motions
of the planets, and Laurie set to work: 'I immersed myself in
the project, read books by and about Kepler, re-researched
the astronomical data because I needed much higher-precision
numbers than John and Willie provided, wrote the software,
configured the hardware,[†] tried a variety of possible ways of

---

[*] The square of the orbital period of a planet is proportional to the
cube of the semi-major axis of its orbit. Duh.
Kepler was essentially tweaking and improving the model of
Copernicus, explaining and predicting the eccentricities of planetary
movements through slightly elliptical, rather than perfectly circular,
orbits.
[†] On a computer running the GROOVE (Generating Realtime
Operations On Voltage-controlled Equipment) hybrid system.
Laurie says: 'GROOVE was essentially a library of FORTRAN IV
and assembly language subroutines that could be called by the
FORTRAN programmes each of us individual users would write
for our various individual purposes. Those routines provided to the

interpreting what Kepler had written, and then, over several months, I generated and recorded the sounds the way I felt sounded most musical of the possibilities I had tried, and gave Willie a high-quality recording.'

In this planetary polyphony Mercury is represented by the highest note, Saturn the lowest. Laurie used only the six planets that were known during Kepler's lifetime. She told me: 'I'd rather not get too technical here, but I programmed each planet as a periodic function using the FORTRAN IV language on a DDP-224 mid-1960s computer. Once I had decided on the timbre and treatment of the sounds and a rate at which time (planetary motion) would proceed relative to listening time, I chose the astronomical date of 0 January 1977 as a starting point and let the program run off 100 years of sound at the rate of 20 seconds per Earth year.'

Job done, the Yale professors put out a press release, and the 'Music of the Spheres' won some extended coverage in the *New York Times* (22 March 1977), which is presumably where Carl or one of the team heard about it before then calling Bell Labs. Laurie was immediately 'excited and honoured' that her piece was being considered. And she set to work, eventually submitting a new version of the piece on a quarter-inch, stereo, open-reel audio tape. 'Carl Sagan's team only included on the Voyager record a very small part of [my] recording. There was quite a lot of other material to pack onto that gold disc. I felt honoured to be placed at the opening of the "Sounds of Earth".'

She thinks that Carl and the team may have hoped that the audible trajectories of the six planets might ultimately act as an identifier of our home solar system were intelligent extraterrestrials ever to find it: 'There turn out to be many stars with planets in the habitable zone, but it's extremely

computer's hardware input and output channels and provided some basic logic and arithmetical functions, such as (notably for this work) a then-relatively-complex periodic function generator. So GROOVE might now be thought of as being a software development environment rather than an application programme.'

unlikely that any will have six that move in these same ratios. This is a sonic picture of the inner six planets of Voyager's unique home solar system – sort of a cosmic fingerprint.'

Laurie's right that they did only use a fairly short excerpt. It's hardly a commercial toe-tapping unit shifter – frankly, it would sit very happily in the John Carpenter songbook, with its cold, detached menace – but it is an interesting mathematical experiment, and a thought-provoking way to kick off the 'Sounds of Earth' segment of the record. It's surprising that it isn't more widely known.

'I never considered it a musical composition of my own,' says Laurie. 'I didn't originate this work. It's a realisation, a sort of a long-delayed quasi-performance of Kepler's *Harmonices Mundi*, his idea that the movements of the planets could be heard as music only by the ear of God ... It had remained theoretical, a mere concept. Then it became possible to render it accurately into audible sound with the development of digital computers and their evolution into sound-creation tools. It was, in some ways, as much of an honour as inclusion in the Voyager mission simply to be the first person to ever hear what Kepler described and then to share with other ears what Kepler never got to hear himself, what he wrote so long ago, hoping that someone would someday be able to listen to the sounds he theorised.'

Laurie sits within a line of pioneers, minimalists and inventors, from Bebe and Louis Barron (*The Forbidden Planet* soundtrack) to Delia Derbyshire, Dick Mills and John Baker (BBC Radiophonic Workshop), to Éliane Radigue, Wendy Carlos, Pauline Oliveros, Alvin Lucier, Suzanne Ciani, Phillip Glass, Terry Riley, Steve Reich, Tangerine Dream, Robert Moog, the whole krautrock crowd, Brian Eno... You'll find her work in compilations celebrating pioneers of the time – *The Early Gurus of Electronic Music* features 1974's 'Appalachian Grove', for example. And alongside compositions, Laurie also developed the 'Music Mouse' in 1986, an early 'intelligent-instrument' designed for home computers, allowing diverse ways of generating melodies and harmonies from gestures.

I guess what I'm saying is, if Brian Eno or Philip Glass had worked on 'Music of the Spheres', I feel we'd all know about it. But they didn't. Laurie Spiegel did, and I think she rules. So go check out 'Music of the Spheres', and once you've done that, go listen to Laurie's *The Expanding Universe* album. It's fabulous.

In any event, her boss Max Mathews was delighted that something created in his department would be on board a spacecraft, and he taped up an article about it from the in-house *Bell Labs Record* on the wall of his workspace.

'I am asked about it intermittently,' she says. 'Usually just when there is news about the mission. That's less and less frequently now they've reached the outer edges of the heliosphere. It still feels like quite an honour, and it's even still hard to believe. Something that I, a mere tiny being on the Earth's surface, made is among the very first human creations to travel beyond the solar system. It's hard to even wrap my mind around that having happened.'

★★★

The Voyagers aren't ever going to land anywhere. Assuming they don't get hit by anything, they will drift in a vast orbit around the Milky Way. They'll be forever in deep space. Therefore, any civilisation with the technology to traverse deep space and find either of the Voyagers should also have no problem in turning an audio signal into a video image. That was the idea at least.

If one of the Voyager records is ever found and decoded, the first image to appear will be a circle. This was another suggestion that originated from Philip Morrison at MIT. He felt a simple geometric shape would be a good way to start the picture sequence – a handy primer for the aliens to test their hardware.

In the end, the team chose a circle as their calibration shape because, unlike a square, it could be used to confirm the correct ratios of height and width in the picture raster* at the same time. Plus, it was designed to neatly work with what Frank had in mind for the record cover.

Frank had been kicking around a number of ideas for the metal casing of the record. They needed something that indicated to an alien being that what was behind the metal box was significant, and that would explain how to use and interpret the contents. We'll explore the cover in more detail nearer launch, but in the meantime, the metal case would be adorned with Frank's Pioneer pulsar map; a schematic of a hydrogen atom; visual instructions about how to play the record, showing stylus, grooves, spin rate and direction; and a sequence of drawings indicating how photographs should be reconstructed from the audio signals.† Next follow diagrams designed to communicate the correct aspect ratio of each image, and then a final picture of a circle. So, putting ourselves in the consciousness of some alien technician many thousands or millions of years hence, once a perfect circle appears on their screen, they know they're doing it right, and they will note that it corresponds to the image of the circle on the casing of the artefact.

Next in the picture sequence comes the pulsar map, showing our position in relation to pulsar stars. The map is a spiky affair, as each pulsar is represented as a series of binary lines indicating the rate of the pulsar, coming out from a central black circle. They wanted to include the whole map, but soon realised that the nature of their weird TV-signal to LP-record format wouldn't result in clear enough resolution to render the binary code of the map. But Frank was definitely committed to using the map on the record's cover, and so was keen to get it on the record in some form – it would again be another clue for the recipients that they were doing it right.

---

* The dot-matrix data structure of a rectangular grid of pixels.
† The idea was that these sound-converted video signals were different enough from all other sounds on the record, that it would be guessed that they stored information besides sound. Plus there were to be repeated tones at the start of each.

The solution was to include a photograph showing just a partial section of the pulsar map, enough to be recognisable, with an inserted picture of the Andromeda galaxy. It was thought this could also serve as a useful signpost or reference point – the shape of the galaxy, which itself will change over time, could provide a clue about very roughly when the record came from.

None of this, however, was getting them any closer to the aforementioned problem of scales of distance, weight and time. They could use a dot to indicate one, but one what? They could define binary notation, then spell out 100 ... but 100 what? They needed some kind of mathematical key, some kind of unit.

The 'hydrogen line' is the electromagnetic radiation spectral line that is created by a change in the energy state of neutral hydrogen atoms. I don't really know what that means either, but as it's universal, is observed frequently in radio astronomy, and is exactly 21cm in length, it seemed a logical fit for the problem. It gives the recipient who understands what Frank is getting at, a unit of measurement of 21cm.*

In the Golden Record version of binary, a vertical line stands for '1', a horizontal line for '0'. Images three and four in the sequence are the keys to the city – a series of mathematical and physical unit definitions, simple diagrams, written out on white paper. Image three (directly following the calibration circle and the partial pulsar map) starts with a single dot, which equals '|' (binary) or '1' (base 10). Meanwhile two dots equals '|-' (binary) or '2' (base 10). It goes on, building up to more complex forms, showing that one-third plus one-fifth is equal to eight fifteenths, for example. They wanted enough sums and examples of mathematical form and notation that a recipient could hopefully grasp what the symbols meant, but also have back-ups to test their hypothesis.

Image four is the physical unit conversion table. It starts with a diagram of the hydrogen atom undergoing a change of

---

* The record cover would include binary notation of the 21cm emission of neutral atomic hydrogen.

energy states – a change that emits radiation at a particular frequency. This is shown essentially by two circles, with a few dots and lines to indicate transition, and a wavy line. Its mass is shown as 1M, and then it gives the frequency of the emitted radiation (1t) and the wavelength (1L). So hydrogen has given the alien a unit of mass (M), time (t) and length (L). And the rest of image number four takes these units and ramps them up to usable figures, translates them for use in the coming picture sequence.

The hydrogen atom has provided us with a universal unit of time. A single 't' is the time it takes the hydrogen atom to do its stuff. And, because this is such a tiny fraction of a second, below the atom symbols we have this unit converted up to one second. It's basically saying that a boat-load of these tiny 't's equals a second.* Below that we learn that 86,400 of these 's' thingies equals a 'd' – a day. Then we discover that 365 days equals 1 'y'.

Next comes the hydrogen length, L, converted to a centimetre. So as the hydrogen line is 21cm, this is a nice easy one: $1/21$L equals one centimetre, which is then expressed as a metre and a kilometre.

Finally the picture repeats this trick with the 'M' – showing how many hydrogen 'M's would equal a gram, then a kilogram. They then invented a brand-new unit of measurement expressed as an 'e'. An 'e' was one Earth's mass. So, having defined one gram, they defined one 'e' as six times 10 to the power of 27 grams.

Jon Lomberg also wrote about this, summing up all the above rather neatly: 'A diagram showing a stylised hydrogen atom. Units of mass, length, and time have been defined in terms of the hyperfine transition of hydrogen – how much hydrogen weighs (M), how fast the transition happens (t), and the wavelength of the radiation produced in the transition.'

I'll admit, because I'm not a physicist, chemist, mathematician or radio astronomer, my brain becomes a little

---

* In printed form this reads one and 42 hundredths, times 10 to the power nine 't's, equals an 's' – a second.

foggy with all this stuff. The hydrogen key has attracted criticism over the years for its complexity, but it's important to remember that the finders of Voyager 1 and 2 are going to be space-travellers, meaning they should have the tools to interpret the symbols correctly. The undoubted genius of all this is that – using a circle, some small lines, a few dots and a wavy line – Frank gave them a key. From these hydrogen units they could now express length, mass and even time on the photographs that were to follow.

★★★

The suggestion to include a series of photographs of bodies from our solar system came from Canadian astrophysicist A.G.W. Cameron. The rendering of three-dimensional objects in two dimensions might be an unfamiliar concept to aliens, so it was felt that some pictures of planets and a star was a nice friendly way to begin the image sequence. They provide some common ground, objects that would be familiar to space-faring aliens.

Having given the alien record collector a unit of measurement, Frank next cracked on with a diagram of the solar system. Remember the recipients who had followed the clues, and understood that this earthling centimetre was $1/21$L of the hydrogen line, now understood kilometres. They also had a time stamp derived from the frequency of the hydrogen wavelength, and a rough measurement of mass measured in a new unit – one 'e', or 'earth'. So next Frank drew up a complete guide to the planets of our solar system, showing diameter and mass. It's a very simple diagram, the kind you might find in any astronomy textbook, but because of the relatively poor TV-quality resolution in this record format, they had to be spread across two images. The first shows the Sun (whose mass is expressed as 333,000 earths), and the four inner planets, all with diameter, distance from the Sun, mass and the rate of a single rotation. The second shows Jupiter, the rest of the gas giants, and of course Pluto (this was long before its declassification).

The first photograph to follow the solar system diagrams is a composite from the Hale Observatory, showing the Sun through various filters and revealing the surface of the Sun, its sunspots and more. These are followed by images showing Mercury's acned face, Mars, Jupiter and Earth, taken from space. Again, the images include the mass and diameter, thereby tying in with the solar system chart information from just before.

You may be asking yourself: 'This is all very well, but exactly how did they get photographs onto a metal record?' It's a good question. It was Frank's job to figure out exactly how to do it. He knew from the beginning it was possible in theory – no problem at all with the *theory* – but the practical how-to was a little more tricky. For a start, no one was doing it. There was really no need to transfer pictures to an ordinary vinyl disc, let alone a metal one.

First they had to find some kind of hardware that could convert picture into sound. For those of a certain age, the old dial-up internet sound might be as good a sound as any to have in mind while thinking about this. The sound of data. More specifically, they needed something that would convert television/video pictures to the lower-frequency signals that could successfully be transferred to a record.

The man who solved the problem was Valentin Boriakoff, another star running back of the Voyager story. Val, as he was more widely known, was born October 1938 in Argentina. He graduated from the University of Buenos Aires in 1963, serving as an electronics engineer at the Argentine Radio Astronomy Institute before migrating to the US, where he first worked at Green Bank, entering Cornell's graduate programme in 1967, receiving his PhD in 1973. In 1977 he was a research associate at the National Astronomy and Ionosphere Center. Indeed, according to Frank, he was one of their star designers. So Frank dropped the problem in Val's lap, while the rest of the team carried on finding pictures, working on the diagrams and cover designs.

It was while flicking through an electronics directory that Val came across the name of a small company called Colorado Video. Not only had they developed a machine that could transfer video image to sound just as required (a 321 Video Analyzer), but they had built a computer around the machine to run it. Frank writes in *Murmurs* how Colorado Video had been working on the technology, convinced that someday people would want to send pictures via telephone wires.[*] Anyway, it all looked perfectly possible. Not only that, but it seemed as if each picture would take up much less time on the record than the original estimates, meaning the number of images could go up yet again.

'Nobody had ever done it before,' says Jon. 'This is not a laser disc or a digital disc. Nobody had ever put pictures on an LP. Frank came up with the idea of using video signals … and also solved the problem of how do you show colour. How do you provide any way that the aliens can reassemble the pictures in accurate colour? Another conceptually very, very difficult challenge. He came up with a solution that in the intervening years a number of scientists, including cognitive psychologists who study vision, have looked at and concluded that it was actually a pretty smart way to do it.'

This is how they tackled colour photographs. Within the fabric of the final record, if you popped it on your own turntable at home and played the picture sequence, the black-and-white images would each come as a single burst of data – a single burst of sound – while the colour images would occur in triplets – three bursts in a row. And a little like old analogue TVs would have those three coloured dots at the centre when you switched them off, showing the constituent parts of the picture, so this record would 'play' the amounts of red, blue and green for each image.

Now for a start, the design team knew that any non-human intelligence might well have eyes or other sensory tools that operate in a completely different way to our own. They might interact with light in a different way, see in a different way.

---

[*] Ha! Ridiculous!

But, however an alien 'saw', they would need some kind of help to interpret and differentiate between the single-burst black-and-white tracks and the triple-burst colour tracks. So the first colour image (image #8 in the finished sequence) is the solar spectrum of our sun, a rainbow strip of light, broken up by hundreds of fine dark lines, known as absorption lines.

Frank reasoned that a spectrum, with absorption lines, should be familiar to any beings that practise astronomy. It's one way you can analyse a distant star. The solar spectrum is like the tell of a poker player, the distinctive giss of a bird, the thermometer you pop under the star's tongue, revealing not only its surface temperature but also composite elements.

So, someone playing the record for the first time, might 'download' this burst of information. While they might not notice at first that it was part of a triple-burst of information intended to be put together to form colour, they might – even in black and white – recognise the distinctive absorption lines of a spectrum. And someone familiar with absorption lines would also be familiar with the absorption line patterns of particular types of star, and this pattern would be familiar as a G2 star (which ours is). So that would make them think, 'Hey, this one should be in colour!' If they can then reconstruct the record's image of the G2-type rainbow correctly, it can serve as the calibration tool for the rest of the colour images in the sequence.

With all that in mind, you'd think that if you wandered through the Cornell Astronomy Department in the mid to late 1970s, you'd easily find multiple diagrams or colour cards of the Sun's solar spectrum. But the libraries and bookshelves of Cornell failed them. So Val and another NAIC bod named Dan Mitler created their own, photographing the Sun through a prism.

The images kicked off with calibration circle, hydrogen-defined mathematical units, diagrams and pictures of the planets. Then, soon after a spectrum to help aliens correctly 'download' the distinctive triple-burst colour shots, comes the first full-colour photograph in the sequence. It's Earth, seen from space, showing the entire orb with the diameter noted in kilometres. This is followed by a second picture of Earth, this time much closer, with visible features including Egypt, the Red Sea, the Sinai Peninsula and the Nile. This means the sequence shows pictures of the Sun, planets, the Earth, and then another, closer shot, as if the camera is descending to Earth, with atomic symbols for water, oxygen and carbon.

We don't know if an as-yet-inconceivable type of biochemistry exists elsewhere in the universe, but the team wanted to share knowledge of the chemical building blocks of our planet. This was communicated by a series of three images that together showed the structure of DNA, but with some unique peculiarities. Again, the picture team found that extant diagrams didn't *quite* illustrate what they wanted. So Frank and Jon, with input from Cornell biochemist Dr Stuart Edelstein, created some new ones. These included schematics of the five atoms that make up DNA, each given letter symbols and an atomic number.

They were so careful to avoid ambiguity at all times. Worried that the 'h's and 'n's used in the DNA diagrams appeared too similar, they gave their 'h' fonts a unique little flick at the top, so every 'h' looks like it's got a baseball cap on backwards. They had defined carbon with 'C', but for the more complex DNA spirals they also needed to intimate 'cytosine' – another 'c'. Jon pointed out this problem to Frank, who suggested they simply spell cytosine with an 's'. And so a version of the famous double-helix diagram exists with a uniquely tweaked lower-case 'h', and cytosine indicated by an 's'. And because the resolution* of this section of the diagrams was at its minimum limit, they repeated the spiral section of the double helix in image #16, and then again in outline alongside a close-up photograph of cell division in image #17.

★★★

---

* By now the team had a video camera and small TV set up in the Cornell workroom so they could view the pictures live in the resolution in which they would be recorded.

'The Family of Man' was an ambitious and influential photography exhibition that opened at the Museum of Modern Art in New York in January 1955. It was a sensation that went on to tour the world, appearing in 37 countries over a period of eight years. 'The Family of Man' had a lot in common with the Voyager's picture sequence – it was a series of themed images designed to be consumed as photo essays. So it is perhaps unsurprising that several pictures from the exhibition (and its accompanying bestselling book) would be included in the Voyager picture sequence.

The original exhibition told the story of different aspects of humanity. The section on childbirth included an image by American photographer Wayne Miller, showing a baby at the very instant of birth. Wayne Miller, who died in 2013, was a Magnum[*] photographer, most associated with his images of the Second World War (he was one of the first photographers to visit Hiroshima) and Chicago's South Side. Voyager Image #22, Birth, is an amazing photograph. You can't see the mother at all – Joan Miller is completely obscured by sheets – but you see a baby all right, held by one leg by a doctor dressed in white, looking down through round spectacles. It captures the instant and motion of birth, the baby still glistening with amniotic fluid, umbilical cord un-severed, taken on 19 September 1946.

When Jon called to secure permission, Wayne was out of town. He explained what was going on, why they wanted the photograph, and was met with gasps down the phone. It turned out he was talking to the baby, Wayne's son David Miller. He guaranteed that even without talking to his father, they could use the image. They also learned that the bespectacled physician holding the baby was Wayne's own father, Harold Wayne Miller, an obstetrician at St Luke's Hospital in Chicago.

[*] An international photographic co-operative owned by its photographer-members.

The Voyager version of this three-generation creation comes with some digits printed along the bottom. The digits read: '22982400s', and below that '266d'. It tells the viewer that what they are seeing, the birth of a child, happens some 22,982,400 seconds or 266 days after conception. And this follows a series of photographs telling the story of human reproduction, each with a time stamp. Photograph #28,[*] for example, was taken by Swedish photographer Lennart Nilsson, famous for pioneering photographs of conception and fertilised ova. His picture shows two spermatozoa just as they are about to reach a human egg. This is preceded in the Voyager sequence by a simplified diagram version of the same image by Jon Lomberg, this time showing one of the two sperms actually reaching the egg – in other words, the exact moment of conception. And next to this, alongside the symbols for male and female and the scale, is the first time stamp: zero seconds.

Next comes two of Nilsson's photographs in one. Also taken from *A Child Is Born* (first published in 1965), they show the thickening of the membrane around the egg, and the first cell division. These two are time-stamped – the thickened membrane is one second into the human reproductive cycle, and the cell division is 43,200 seconds in.

There follows another silhouette by Jon showing two foetuses at different stages of development, again with the time since conception and their approximate height (3,456,000 seconds, 2.5cm), before a photograph of a foetus after about 60 days. Then we come to image #32, another diagram by Jon, this time showing the silhouette of an adult male and female. There's no hint of reproductive organs, but there is a male and

---

[*] Another person who's often left out of the Voyager story is Roscoe Barham. He was working at the National Academy of Sciences in Washington and, according to Frank, essentially agreed to be the go-between and messenger, driving between the George Washington University, the National Geographic Society and the airport in order to make sure that the vital slide of an embryo arrived in time.

female symbol, approximate height in centimetres, age in years, and the woman is shown carrying a foetus in her womb.[*]

The final time-stamped entry is image #37. This is another silhouette image – a simplified version of a 1947 photograph (#38) of a typical Midwestern American family, another image drawn from the 'Family of Man' collection. The image, first printed in *Life* magazine, shows a multi-generational group, and in the last of the time-stamp keys, Lomberg's simplified silhouette draws out four of the figures, giving their age and approximate weight. The original photo was by another noted American great in Nina Leen. Jon was in New York, visiting the offices of Time-Life to secure permissions. By coincidence, Nina was there too. Jon told her what they were doing, informing her that they intended to show her photograph to alien beings. Her reaction was positive. However, Jon noticed that she not only seemed happy, but strangely unfazed and unsurprised about the whole thing. Then she explained to Jon that she'd been in contact with aliens for ages. She said they knew all about the Voyager record project, and were pleased with the idea.

★★★

The team had introduced a scale, our solar system and the human race – starting with birth, anatomy and ending with the family group photograph (#38). Now they wanted to introduce Earth in more detail, to give the aliens a view of our terrestrial geology and then our biosphere. We'd already had one picture of the Earth from space in full colour, the first of the full-colour snaps after the calibration spectrum, and now there followed a series of photos in black and white.

The Earth sequence starts with a diagram of continental drift – adapted from Carl's design for the LAGEOS plaque.

---

[*] This diagram also had a photographic sister: an asexual image of a naked couple standing hand in hand, the woman visibly pregnant. But as we know from the Pioneers, NASA was not keen on nakedness. And by the time Voyager came around, little had changed.

This showed the arrangement of the continents many years ago, how they appear now, and how they would look several million years hence. Jon adapted it slightly, giving a time stamp in years, and adding a black silhouette of a single open hand next to our current configuration of continents, as if to say: 'Hi! This is us just here.' Next comes a diagram of Earth's structure, again drawn up by Jon with help from Steven Soter, who worked with Wendy in the office across the hall from Carl. It shows the outline and dimensions of Earth's subterranean layers, the core and mantle, and notes the most abundant elements found on Earth.

All of this is the slightly dry primer to the fun stuff: picture-postcard Earth, a sequence of photographs showing our myriad landscapes. There's a coastal shot of Cape Neddick in Maine, showing wind and surf and a lighthouse; there's a breathtaking Ansel Adams shot of Snake River in Grand Teton National Park, Wyoming; there's Monument Valley; there's a lone horseman crossing some sand dunes.

And then there's image #41, my favourite. 'Heron Island on the Great Barrier Reef' was not taken by Ansel Adams or by any other celebrated photography great. It was taken by an astronomer named Jay M. Pasachoff, who told me all about it.

In March 1974 he and his wife were married. They had a kind of first-stage honeymoon in New York City, then, in June of that year, went for a second, full-blown honeymoon in Australia – first to witness a total solar eclipse in Albany, Western Australia, then to a radio-telescope observing session at the Parkes Radio Observatory in eastern Australia, and finally to Heron Island in the Great Barrier Reef. 'We flew out from Brisbane in a helicopter,' Jay says, 'and the photo on the Voyagers is one I took with my Nikon, aloft in the helicopter.'

Jay knew Carl from his student days, studying at Harvard (he graduated in 1963), where Carl was an assistant professor of astronomy. By the late 1970s, Jay was in Cornell, where he would mingle with Carl from time to time. Jay remembers Jon Lomberg, but he suspects it was actually Wendy Gradison who first approached him and selected two of his images: the Heron Island picture, and another of the birth of his eldest daughter.

'NASA censored the latter, no doubt as too graphic,' Jay told me. 'My daughter joked much later on that she was relieved that any aliens who came to Earth wouldn't see that photo and say: "Take me to your Pasachoffs."'

Jay has been researching the atmosphere of Pluto for the last 15 years, studying it when it occults[*] distant stars. In 2017 he was part of a NASA mission to Argentina to try to pinpoint the position, again by occultation, of the next target for NASA's New Horizons spacecraft. The vessel, which flew by Pluto in 2015, is due to reach 2014 MU69[†] on 1 January 2019.

'It is wonderful that I have a photograph on the Voyager spacecraft,' he says. 'It is fun, too, to have my name adjacent in a list of photos to that of Ansel Adams, one of the greatest photographers of all time.'[‡]

Next the team wanted to introduce vegetation, so they included a photograph of a forest, a close-up of a strawberry leaf, a colour photograph showing a deciduous tree with a woman sweeping fallen leaves from beneath it. There follows a series of photo composites – a colour picture of the sequoia tree, covered in snow, with inserted a close-up image of a single snowflake, showing the crystalline structure that, it was felt, might well be familiar to an alien being. Then comes a second colour image of a naked, early springtime tree, with a 14m height scale added, and surrounded by daffodils. Added to this is a close-up picture of a daffodil.

★★★

---

[*] Studying a celestial object as it crosses the line of sight between the observer and another celestial object.

[†] A trans-Neptunian object from the Kuiper belt located in the outermost regions of the solar system and discovered by astronomers using the Hubble Space Telescope on 26 June 2014.

[‡] Jay receives requests for copyright permission from time to time. One came from artist and director Steve McQueen for a 2002 piece put on at Musée d'Art Moderne de la Ville de Paris. The film was called *Once upon a Time*, a slow projection of 118 successive stills, using images from the Voyager probe, set to a soundtrack of 'interwoven snippets of gibberish'.

One of the most thorny and tedious elements of the Voyager record would be sorting out copyright. NASA was nothing if not a 'by the book' kind of organisation, and although there was no specific law covering the alien consumption of human-made music and images, they still had to seek permission. When in May 1977 a letter arrived at photographer Stephen Dalton's house in Sussex, asking whether one of his pictures of an insect in flight could be sent into space, he assumed it was a joke. 'I thought the letter was a spoof. I have a few friends who regularly play these sorts of practical jokes on me,' he says.

Stephen pioneered techniques to capture insects on film in a way they had never been seen before. 'Insects are the most successful terrestrial vertebrates in the world almost entirely because they can fly,' he says. 'Yet prior to 1970 there were no photographs of them actually flying – all were sitting or crawling. After finding out why, I decided to record them in full free flight with no compromise in critical detail, in full colour and in their natural settings. It took two years to develop suitable equipment and techniques. I cannot remember anything about the day the photograph was taken beyond saying that it is a very early flight photograph, and about my least favourite photograph of a flying insect. Why NASA chose this one is a complete mystery as I had plenty of other more impressive ones to choose from!'

I recommend tracking down a copy so you can have a look yourself. In full colour and high resolution, it's a crystal-clear snapshot of a rather leggy-looking wasp on the wing. It's officially image #51 in the sequence, appearing just after a tree in bloom with a daffodil close-up, and was used in black and white.* But it looks much better in colour.

'Oddly enough, as the years go by, I feel increasingly privileged at having been asked to have one of my photographs included on board the Voyagers, sitting alongside phenomenal works of art by geniuses like Bach! I now regard that picture

---

* It's reproduced twice in *Murmurs*, and in one place it is reproduced with the insect flying upside down.

with a certain amount of awe which I never did when I took it. To think that these little "space rockets" together with their records of Earth have now left the solar system and have now entered outer space – and will probably outlast the life of our planet – is difficult to get one's head round. It is, in my opinion, a more impressive feat than anything else humans have achieved, including putting man on the moon. Apart, perhaps, from the discovery of DNA.'

Image #52, following Stephen's wasp in the biosphere sequence, is another diagram by Jon Lomberg. It's a pretty simple diagram, showing a rough evolutionary sequence of life, from sea to land – from a shark-like outline, through reptiles and birds, to deer. It's the kind of diagram you might find in any natural-history book, and was adapted from the book *Life: Cells, Organisms, Populations* by E.O. Wilson. It seems quite a simplistic sketch at first glance, and it begs the question – why not just reproduce a diagram from somewhere else? The reason Jon chose to create his own was so that it could be tailored to work with other images in the sequence. With the exception of the first drawing (of a shark) and the third (a kind of fish with feet), all the others mirror animals that appear elsewhere in the image sequence – so the bird is an outline of the eagle that appears in image #58, and the grazing deer was traced from an animal being stalked by two bushmen hunters in photo #62.

Jon's diagram is also notable for containing Voyager Golden Record's only proper joke. Just to the right of the deer stand two human figures. At first glance they look identical to the human figures from the old Pioneer plaques, expertly drawn by Linda Sagan and with the man's arm raised in greeting. This time, however, it's the female figure that is waving to camera. Jon says: 'There were two criticisms of the Pioneer plaques: one that it was smut, the other that it was sexist. I don't think either is true. I didn't really discuss it with Carl, but I just decided to do it myself, to show that we had heard the criticism, we were sensitive to it, and that … well, you had a 50/50 chance. One of them has to be raising their hand at the time; we'll let the woman raise her hand. … This would be the one time where my thought was not the alien audience, but the human audience. Definitely.'

The life sequence was now in full swing. After Stephen Dalton's flying insect and Lomberg's evolutionary diagram comes a beautiful picture showing the cross section of a seashell,* then some jumping dolphins, then a school of fish, then a tiny tree toad in a human hand. The team wanted pictures of wildlife that also included human figures. This would show a number of things – that we share our planet, that we hunt, exploit and study other species – and it would give the pictures a sense of scale. So we have a figure swimming through a school of fish,† a man measuring the tail of a dead crocodile on its back, a pair of bushmen hunters stalking a deer (another from the 'Family of Man' collection), and a picture of Jane Goodall filming the behaviour of chimpanzees. This last example was taken by her mother, novelist Vanne Morris-Goodall, who, according to Jon writing in *Murmurs of Earth*, was delighted that her photograph and her daughter's work would commune with the cosmos.

Jon writes that he had a very zen approach to selecting the animals. With so many hundreds of thousands of species of animals to choose from, it was too overwhelming to decide beforehand which ones would best represent the fauna of Earth. He didn't necessarily want to follow the most obvious, expected route of using all the 'charismatic vertebrates that zoos use as poster boys' but that are only a tiny fraction of Earth's biomass. 'There were very few specific images or animals we searched for in particular,' he writes. 'We were Zen archers. The photos would find their way to us. We would simply be receptive to the blizzard of images passing across my desk ... Some example

---

* It's a beautiful photo and actually one of very few in the final list that Jon chose purely for aesthetic reasons. Both Frank and Carl felt it was too open to misinterpretation.

† Taken by David Doubilet in the Red Sea off Na'ama Bay in Sinai. According to *Murmurs*, Frank, Carl and Jon were all keen scuba divers, and wanted a picture that showed humans using breathing apparatus underwater.

from each taxonomic order, if not each family, seemed required. Here, as elsewhere, we were limited by what we could find and secure permission to use in the limited time available. The presence of a human in the photograph was a plus, both to provide scale and to show our interest in studying the beings with whom we share our planet.'

Jon says: 'If you wanted to show a picture, let's say, of a hammer – just showing a picture of a hammer against a white background or a grey background with nothing else in the picture is the clearest way to show a picture of a hammer, but it doesn't show you what it's used for. So maybe you want a picture of a man holding a hammer, but then maybe you want a picture of a man holding a hammer actually nailing something. Then you want to see the actual building, so you see the context of what he's nailing. And then you want to show that there are other people working there as well. So you start looking for your picture of a hammer and it starts getting more and more complicated. Now there's so much in the picture that you don't even notice the hammer and it's just a confusing bunch of objects. So I was looking for the ideal picture that had a lot of information in it but where the clarity of the presentation was also there.

'The other thing I started to do was to try to build in recurrent motifs. In other words, there were certain things that people had said are so important. You want to make sure that they get it and the best way to do that is to say it more than once. So an example might be the human hand. I really wanted to make sure that there were lots of pictures of hands doing all the things that hands can do. Similarly if there were other objects that could appear in different contexts in different pictures. Then if one picture wasn't understood, the fact that another one was could let them go back to the first one and say: "Okay, now I see. This makes a little more sense." So I was looking for recurrent images that would help link a story, for example in the sequence of how we eat. There's a picture of a fish in the water, then fishing nets, then a number of fish on a grill slowly blackening as they cooked. The fact that you saw the fish in its natural habitat, you saw the means of catching it, you saw it on the grill* – helped make that story clear. Had it been a fish, and then he was cooking lamb chops, you wouldn't have had that connection.'

As they continued to whittle down the picture sequence, they built a Rube Goldberg† rig in their workroom so that they could check the overall look and final resolution of the video-converted photographs. Each candidate picture would be in the format of a slide – either they were supplied as a slide or, if not, HAIC photographer Herman Eckelmann would create a new slide. They would project that slide onto an A3 piece of paper that was taped to the wall and then, using a video camera mounted on a tripod, they would video the image being projected on the wall, and see how it looked on a tiny television screen.

In the film version of this story, now would be the perfect moment for a montage. Let's have some piano-led music, not unlike the jolly theme tune from *Murder She Wrote*. Let's picture Jon, dark hair framing his face, brow furrowed in concentration. Jon looks up at a wall calendar. There's the June deadline, and we see the days as they're crossed off in red magic marker. Wendy walks in, gives a megawatt smile, adds to the teetering pile of books. Frank throws him a *National Geographic* with a post-it note attached. Jon sighs, but opens it up. Now Wendy's seated at the same desk, Jon standing behind her and reading *Gem Cutting*. The clock shows 9a.m., 5p.m., 11p.m. Then Jon is seated back at the desk, rubbing his eyes. He cuts out another photo and adds to the pile. We see Wendy, Shirley, Amahl Shakhashiri (Frank's assistant), Jon and Frank all talking animatedly on the phone. More subjects are ticked off the wall chart. Then the music fades, and we rejoin Jon, a mug of coffee in hand, staring at the wall. The camera moves closer to show us what he's looking at. It's the wall chart. The camera moves closer still. We see that lots of

---

* This photograph came from the Time-Life book *The Cooking of Spain and Portugal*.
† The US equivalent of Heath Robinson.

the subjects now have ticks next to them, but we notice that a couple do not. We read two that don't: 'Waterhole' and 'Eating'.

★★★

The only animal sequence photo they had in mind *before* they started looking, was animals gathered at a waterhole. This was an idea Frank had right back at the beginning – a single photograph that would show a range of species at a single point, drinking. The scene would emphasise the importance of water to life here, and would show several non-human species co-existing. There was also an obscure pun built in to the waterhole picture, an in-joke for people who shared Frank's specialisations and interests. You see, radio astronomers had found that a certain portion of the radio spectrum seemed to be the quietest and most effective in terms of transmission through planetary atmospheres and interstellar space. This band of frequencies had been dubbed the 'waterhole' as they are sandwiched between the frequencies of the emission of hydrogen at one end and hydroxyl radical at the other – which combine to make water. So, just as animals gather at the waterhole, intelligent beings might meet at the 'waterhole' region in the radio spectrum.

The best, indeed the only, scene at a waterhole that the team were able to find appeared in a brochure published by the South African Tourist Board. They gave their permission easily and willingly. Jon writes that back in 1977, as South Africa was still a pariah state for its institutional racism, they did wonder if it was sufficient reason to eliminate the photograph from consideration. But with no alternative coming to the fore, they decided to include it.

The team also wanted something that clearly illustrated the way humans eat and drink. There are eight images in the sequence that together illustrate food, agriculture and eating. Image #75, for example, is a *National Geographic* shot of three enormous harvesters rumbling over a cotton field. Then comes image #76, 'the Grape Picker', captured by influential

Australian photographer David Moore. 'Grape Picker' isn't Moore's most interesting or aesthetically pleasing work, but it does show a human figure, his forearm covered in sweat and mud, clearly pulling a vine of grapes towards his face and pressing them to his mouth. This is followed in the sequence by a picture of a woman in a supermarket. She is also eating grapes, standing in front of a banked wall of fruit and vegetables. The idea was to connect these two images, both showing humans eating grapes, and to illustrate that not all humans were hunters, gatherers or farmers – that we also buy our food from markets.

The woman is actually Wendy. It was Frank's idea that they should have some kind of image of a market, so rather than spend more time looking for a picture, five of them – Frank, photographer Herman Eckelmann, Wendy, Jon and Amahl – trooped down to the produce counter at the Grand Union Supermarket, Cayuga Mall, Ithaca. Jon was kind enough to show me one of the also-ran photographs from this particular session. In it we see Amahl, wearing a white blouse and black skirt, her long dark hair loose around her shoulders. She's standing sideways to the camera, handing a shrink-wrapped pack of grapes to Jon, who's sporting a light, floral-print short-sleeved shirt tucked into what look like grey-blue jeans.

The photograph has various problems. As the grapes are in a shrink-wrapped box, you can't really see what they are. And from the perspective of a life form unfamiliar with grapes, it's not at all clear what the two human figures are doing. They're both looking at each other, with slightly bemused expressions, holding this pack of grapes aloft like some rare, valuable, god-like or possibly dangerous item. You can certainly see why this *didn't* make it aboard Voyager. It doesn't answer any questions at all, it only poses more of them. What on earth are they doing? Is this some kind of ceremony?

The final image that was chosen, of Wendy eating a grape, is much better. She is clearly eating a grape, and she is clearly standing in a place where other foodstuffs seem to be available. Wendy also looks slightly uncomfortable, but that may have been because of the odd looks their escapades were provoking. According to the account in *Murmurs*, this strange band – posing and being photographed with food, then returning the food to the shelves – was attracting lots of attention, including from the store manager, who angrily enquired what the hell was going on. Frank did the talking, explaining that there was nothing to see here; they were just taking a photograph of a woman eating a grape so they could send the photograph to the stars in a metal record.

The session ended. They returned most of the food to the shelves, paid for the grapes and returned to Cornell.

While man-eating-grape-from-vine (#76) and woman-eating-grape-from-supermarket (#77) both showed humans pressing something small and round to their open mouths, it was felt neither properly illustrated the whole story of how we eat and drink. You might think finding an image of humans consuming food and drink would be simple, but nothing seemed to unambiguously communicate the mechanics of human consumption. The only other image in the running was Chinese dinner party (#81). This unremarkable shot was taken from a book called *Chinese Cooking*, from the Time-Life 'Foods of the World' series, chosen because it illustrates how eating is often a social, group activity. And yet even with this example, it's still hard to be sure exactly what's going on. To us humans it's plainly a gathering around a table stacked with food, but to an alien audience, it's not so obvious. For a start, only one person in the photograph is clearly raising something to their mouth, whereas the rest are smiling, talking or chewing.

Amahl Shakhashiri suggested they stage their own image. They briefly discussed a plan to illustrate drinking with three pictures: #1 with a person looking 'unhappy', then #2 holding a glass with water to their mouth, then #3 the glass down, half-full, and a person with a smile on their face. Instead Amahl envisaged one efficient photograph showing three people, one licking, one biting, one drinking. Graduate student George Helou made a suggestion that a person holding a pitcher full of water, and pouring water into their mouth, would tell the story more clearly than a glass of water. 'Moreover, you could estimate something about gravity by measuring the arc of water,' he says.

George grew up in Lebanon and attended college at the American University of Beirut, before leaving for the US. He had been advised by Frank Drake for a couple of years and had taken courses with Carl. He was in his early twenties, and intent on advancing through his graduate studies quick sharp. He told me: 'I'd always been interested in SETI and authored with Frank an important paper on "optimal frequencies to use in SETI". [I was] also very interested in the science of planetary exploration. So when this team was being assembled by Frank and Carl it was natural that I'd be involved, even though I can't recall exactly how it started … This Record exercise was a bit of a distraction from that pursuit, which was about taking my exams and getting started on a thesis, so I tried to limit the time spent on it.'

They gathered in the studio of staff photographer Herman Eckelmann. Amahl took the lead, while Herman stood behind the camera. The three actors were Wendy Gradison, Val Boriakoff and George Helou.

The first photograph didn't turn out too well. The idea was to show someone licking ice cream, to show the human tongue at work, as tongues didn't appear anywhere else in the picture sequence. And while you can see the cone of chocolate ice cream in Wendy's hand, you simply can't see her tongue clearly enough.

Val, dressed in a grey round-necked tank top over a pale green shirt, had been handed a tuna sandwich in white bread. According to Jon, Val hated tuna, which certainly comes across in the first attempt – he is practically scowling at the camera as he bites down. But his face wasn't the problem; the angle and colour of the sandwich meant you couldn't clearly see the bite at all. Meanwhile, George can be seen pouring water into his open mouth from an opaque earthenware jug, but when they came to develop the photograph, they felt that the water could be interpreted as some kind of solid silvery pipe. Remember, although the final image (which you can see at JPL's online Voyager archive) is in colour, within the record this would be one of the black-and-white, low-resolution images, meaning some of the finer details would be harder to discern. They tried again.

By the time of the final shot, Val is dressed in a simple white shirt, George has on a fetching blue, red and white striped number, and Wendy this time is shown perfectly side on, clearly licking the tip of the chocolate* ice cream with outstretched tongue. Val's sandwich is now on dark rye toast, much more discernible in black and white. And before the photograph was taken, he took a nice clean bite out of the *opposite* side of the sandwich so that the photograph efficiently captures both the event of his biting down on one side of the sandwich as well as the after-effects of his biting down on the far side of the sandwich. George meanwhile is now grasping a clear glass jug,† so while you still have the silvery pipe problem, you can clearly see that the clear jug is full of a liquid, that seems to be being poured directly into his open mouth.

It still took some time to get it right. George had to keep pouring water into his mouth while Herman got the lighting, white balance and focus all right. 'This type of pitcher is ubiquitous in Lebanon, and drinking that way was second nature to me,' says George. Nevertheless, ask George how much water he had to drink before the photo was done and he'll tell you: 'A lot.'

Today George hangs with the infrared posse. Since the early 1980s he has been involved with NASA and European Space Agency infrared astronomy missions that include the still-active Spitzer Space Telescope.‡ Indeed in 2017, right

---

* You can only see it's chocolate in the colour version. Aliens may well assume it's vanilla.

† The pitcher belonged to Amahl.

‡ An infrared space telescope launched in 2003, named after astronomer Lyman Spitzer, who had promoted the concept of space telescopes in the 1940s (www.spitzer.caltech.edu).

around the time I was asking him questions about water being pitched into his open mouth four decades earlier, he was helping spread the word about Spitzer's 'Trappist-1' discovery – the first known system with seven (yes, *seven*) Earth-size planets all around a single star, and including three in so-called 'goldilocks' orbit, meaning they could have conditions right for life. As this system is around a dwarf star only 39.5 light years from the Sun (located in constellation Aquarius), it was heralded as a major step forwards in the search for alien life.

So there you have it: George's fetching shirt, Val's distinctive bite and Wendy's tongue, together forever. Wendy says: 'I believe that my tongue is immortalised in space because I was available and sitting at my desk the day of the photo shoot. Val, George and I were there because three people were needed on short notice. It was pretty funny at the time. Speed, convenience and ease of access were themes for both photos of me. The other photo, eating grapes in a grocery store to set context juxtaposed against the "Man with Grapes" photo was similar. I was available. Understanding, with my non-scientific background, the path and importance of the Voyagers, I always pinch myself that I was able to be a part (albeit small) of this project, and am amused and also awed that there exists a representation of me and my tongue in the cosmos, forever. Although I am CEO of a multi-million-dollar behavioural healthcare company, I still feel having been involved, and being out there in space, is the coolest entry on my résumé.'

George also told me: 'Even then, I understood the record was an opportunity for us humans to look at ourselves and ask what's essential about humanity. Today that assessment hasn't changed. Personally, I feel incredibly lucky to have been at the right place and right time to participate in this remarkable exercise of putting together the record, as an intellectual exercise.' And in his acknowledgements in *Murmurs*, Frank Drake writes that Val, who passed away in 1999, felt his reward for all the help he gave to the Voyager project was being in that picture.

The man who took this photograph, and many others on the record, was Herman Eckelmann. Born in Hoboken in 1925, Eck was a Second World War veteran who had served aboard the aircraft carrier *Franklin D. Roosevelt*. He was married, had a daughter and four sons, and alongside his Cornell work as an electrical engineering research associate and photographer, he was a pastor, founding the Faith Bible Church in Ithaca in 1960. In his acknowledgements Frank pours out gratitude, describing Eck working into the small hours in his dark room, cancelling family arrangements to keep working, taking new original photographs when demanded, and endlessly creating and recreating slide-format photographs for the final sequence. 'Occasionally he had a quick sandwich.' And in 2018 Amahl described Eck as 'indispensable', a go-to photographer, ready to help with close-to-zero notice. Eck passed away in 2001.

★★★

Meanwhile, Carl was still wrestling with the Outer Space Committee's ingenious little catch-22. Yes, they could agree to being recorded saying 'hello', only after they'd had a meeting about being asked to say 'hello', and having agreed to their saying 'hello', they would say 'hello' on a date that would be too late to be recorded saying 'hello'.

Carl again phoned the State Department. Its prime concern was that if they helped Carl in his ambition and did bully some delegates to offer up greetings for the record, they needed guarantees that the delegates' messages would be used. This was all getting very sticky. Carl simply could not make such a guarantee. The very project itself was without guarantee. NASA had the right to refuse the record at the last moment if they found it to be substandard, embarrassing or too full of vulvas. Plus, even if the record was a certainty, Carl would not have wanted to make promises at this stage. There were still too many unknowns. The record had not yet been mixed, the soundscapes not fully cut, the images were not yet encoded. They simply did not know how much editing would be required.

According to Willian Poundstone's Sagan biography, Carl called NASA yet again, and this time found someone useful – a chap called Arnold Frutkin[*] who seemed to know who to talk to at both the State Department and the UN to get the ball rolling. It was Frutkin who persuaded the State Department to take a leap of faith, and it was he who contacted the then-UN Secretary General Kurt Waldheim. Hang on. That name sounds familiar…

---

[*] Frutkin (b.1918) was deputy director of the NASA international programmes office between 1957 and 1978, basically putting him right at the coalface throughout the space race.

CHAPTER SEVEN

Berry v Beatles

*'It's very hard to describe my mental state because I was so full of
adrenaline and so hyped up most of the time. I think that another
advantage of having such a compressed time period was you didn't
have time to be overwhelmed. You just had to make decisions. You
had to keep on going.'*

Jon Lomberg

The most famous track on the Voyager Golden Record is
Chuck Berry's 'Johnny B. Goode'. Exactly who was
responsible for its inclusion is sometimes disputed, but let's be
absolutely clear: it was Ann. Lots of other people were behind
the idea. If the song was a huge oil tanker idling in a harbour,
many other tugboats helped guide it out to sea. But Ann was
the harbour master who first suggested it, and then campaigned
tirelessly for its inclusion.

Written by Berry in 1955, the partially autobiographical
song is about an illiterate country boy who plays a mean
guitar. The name was inspired by Berry's birthplace – he was
born on 18 October 1926 at 2520 Goode Avenue, St Louis –
and partly by his bandmate, piano player Johnnie Johnson.
The year 1955 was a pivotal moment in Berry's career. He
had visited Chicago, signed with Chess, and put out
'Maybellene'. It promptly sold a million copies, becoming the
first in a run of hits from 'Roll Over Beethoven' and 'School
Days' to 'Sweet Little Sixteen' and 'Memphis, Tennessee', all
of them packed with catchy-as-all-hell riffage and story-song
lyrics. 'Johnny B. Goode' was released in 1958, becoming yet
another crossover hit, reaching number 2 on the R&B charts
and number 8 on the pop charts.

Tim told me: 'That was Annie's idea originally. I think I
agreed because you have both the innovator himself, doing
a song about himself, you know about this process. It's a
very unusual nexus in pop music and rock music. So yeah, I

thought it was a great choice.' Besides, Tim knew they were unlikely to have more than one rock piece on the record, so if the team settled on Chuck then that category was all done and dusted – another item ticked off his to-do list. Nevertheless, the song would be contested in daily and nightly debates. For a start, Carl was not a fan. Indeed, the first time Ann played 'Johnny B. Goode' to Carl, his reaction was: 'What? No! I don't like this at all!' When it came to popular performers, or which modern performer should make the record, Carl was firmly in team Dylan or team Beatles. This was just too silly, too throwaway, too lightweight. However, with Ann campaigning, and Tim and others in the group being supportive, his position gradually shifted.

'I've said this many times,' Ann told me. 'When it came to "Johnny B. Goode", I was like Cato saying "Carthage must be destroyed." At one point I think Tim played "Roll Over Beethoven" by Chuck Berry, thinking maybe the classical reference would impress [Carl] more, but he didn't like it either. And I was still saying, "Johnny B. Goode, Johnny B. Goode, Carthage must be destroyed!" And the reason it had to be Chuck Berry was that he was the progenitor, you know that crossover between African, European and American music. And those guitar riffs and those lyrics that went like the lyrics of a novel … I mean, Chuck Berry wrote novels that you could experience in under three minutes.'

Carl may not have liked it either, but he was at least tempted by 'Roll Over Beethoven' – perhaps the fact that it namechecked a classical great also bound for space appealed to him. Bill Nye the Science Guy and CEO of the Planetary Society, was just another young dude taking a class at Cornell around this time. There Sagan put the question to his students: 'Which rock 'n' roll song should we put on? We're considering "Roll Over Beethoven" or "Johnny B. Goode".' According to Nye, everyone shouted: 'No! Professor Sagan, no! Not "Roll Over Beethoven"!' The consensus in the class was that yes, it's good – no argument there – but it just isn't 'Johnny B.

Goode'. Nye was definitely one of the tugboats helping 'Johnny B. Goode' out to sea.

While Carl was slowly coming round to it, Alan Lomax was dead against it. Alan thought it was for teenagers and children. To which Sagan countered that, as the Earth is full of teenagers and children, wasn't it appropriate to have a song on a global interstellar record that represented them too? Tim told me: 'I used to go up and see Alan at his place sometimes, collect pieces of music there. I think he wanted to be more involved in all the deliberations, but he wasn't really very well set up to do that. He hated "Johnny B. Goode", for instance. And he would sometimes dig his heels in. And we didn't really have time for that. We needed to keep our eye on the ball.'

Carl not only came round to 'Johnny B. Goode', but he and Chuck would enjoy, if not a friendship, certainly a warm acquaintanceship. Berry would play at JPL's Voyager fly-by celebrations in years to come, and Carl would sign off a letter to Chuck on his 60th birthday, in October 1986, with the words 'Go Johnny, go.'[*]

'I love the guy. I think he's a genius,' says Ann. 'John Lennon saying to camera, "how do you spell rock 'n' roll? You spell it 'Chuck Berry'." Without him there would be no rock 'n' roll. So I also felt ... vindicated. I learned later that the original lyric for the song had been "coloured boy", but the producers had Chuck change it to "country boy" so it wouldn't alienate a white audience ... He was really a crossover artist pioneering his way into the hearts of this new audience that had not really been paying attention.'

Ann remembers one more critical thing about the Chuck Berry debates. It was a moment she recalls vividly, when Carl looked up at her and said: 'Okay. Chuck Berry. Johnny B. Goode.' She remembers it because there was something in the way he said it, in the way he looked at her. It seemed to Ann that Carl was saying much more.

---

[*] If you want to see the letter, it's online at the Library of Congress Carl Sagan archive.

But let's not get ahead of ourselves. The Beatles were still in the running, remember? And if there was to be only one rock 'n' roll song on the record, surely The Beatles would be chosen over Berry. Right? I mean … it's The *Beatles*.

Abbey Road is the 11th Beatles album, and in a way it's their last proper studio album as the *Let It Be* sessions were recorded prior to *Abbey Road*. It was certainly the last complete Beatles album to which all four members were fully committed. In any event, it's an album, with highlights, lowlights and curios. It's heavily produced, which annoyed some contemporary reviewers, who, alongside plenty of praise, used words like 'gimmicky' (*Times*), 'disposable' (*Life*), 'nothing special' (*NY Times*). Within the album sits what is generally accepted to be George Harrison's best song (before going solo). And it was 'Here Comes The Sun' that some in the Golden Record team were excited about sending.

Now, the often-repeated story is this: they wanted to send 'Here Comes The Sun', they put out feelers, they were told that all four of The Beatles were keen and were happy for the song to go to space but, in the end, the copyright owners, Northern Songs, wanted to charge an extortionate amount ($50,000 for each copy of the record on the two spacecraft), which meant The Beatles missed this particular boat to immortality.

In the 2017 documentary *The Farthest* Frank misremembered this story, telling it as if all four of The Beatles had said 'no'. But that isn't quite right either. It's certainly true that Northern Songs owned the copyright and did want an extortionate amount. Carl writes how there were many other stories like this – great artists and musicians who, because of time or money constraints, could not be included on the record. Carl also writes how they imagined a cartoon printed in a newspaper, showing a group of performers, all staring wistfully up at the Voyager rockets as they took off from Cape Canaveral, mourning their missed opportunity. In any event, because it's The Beatles and because it's a lot of money, this story is told a lot. However, Tim remembers it differently:

'What's often reported is that we were all dying to put a Beatles track on but The Beatles asked for too much money or their record label did … But that's not quite accurate. One of the songs that came through early on was "Here Comes The Sun". I may have proposed it myself, I can't remember who did. I remember we played it at my place and Carl was kind of enchanted, but was concerned about whether it would be released. I was less enchanted because I didn't think the song was going to make it to the record. It's a pleasant enough piece, but it's way down in my opinion … it was way below the standard we were looking for on the record. I went to something at Forest Hills with John Eastman, who was Paul McCartney's attorney … We had a chat about it while we were out there and he said: "You know, I don't see how that could be a problem. But then I don't control the publishing on this particular song."'

Then Tim describes that around this time he also saw George Harrison at a party. Tim remembers that George was being 'all prickly' because of something that had appeared in *Rolling Stone* recently. This wasn't a bad argument, just a spirited exchange of views, and in any event, Tim didn't even mention 'Here Comes The Sun' at the time because he didn't think it would ever make it: 'I was convinced pretty early on that this was never going to make it through competition. There's nothing on the record that's as simple and un-resonant as "Here Comes The Sun". All you really have is a pun, which isn't accurate anyway because we're not sending you the Sun, we're sending you this record. So I stopped pursuing it and went on to other things.'

But while Tim may have abandoned the idea, it seemed someone on the team had not: 'Somebody – Carl maybe, or somebody on the team – apparently called somebody else who came up with some "fuck you" number [to pay for the song's inclusion]. You know, that was probably about the annual income of the average American at that time. And it was just a crude way of saying no. I don't know who said it. But as the piece was never going to be on the record anyhow, I regarded all of that as kind of incidental. So those are the facts, and the conclusion I reached is simply that it's not true that The Beatles aren't on the record because somebody asked for a ton of money. Bob Dylan isn't on the record either. And there are a lot of people who aren't on the record who are just fantastic. You know, I wish they were on the record. But you can't have everybody.'

You certainly can't. No Dylan. Elvis was discussed but discounted. There was no Bob Marley (a fact that still bothers Jon), no Billie Holiday, no Jimi Hendrix, no Aretha Franklin, no Miles Davis and no Rolling Stones. Even Jefferson Starship, who had offered their music for free, weren't in the running.

Nevertheless, Ann is confused by this. She remembers how they were only paying a nominal fee for the rights for most songs, but the labels and performers, including Chuck Berry, were keen, knowing it was a one-time chance, a shot at immortality. She was sure that Tim *had* contacted The Beatles and that all four Beatles said yes, and it was Northern Songs who wanted this huge amount – not only that but a huge amount for each spacecraft.

'It was very funny because we were paying a two-cents-per-spacecraft royalty for "Universal Rights" for all the tracks, which is a phrase that appears in all music contracts and publishing contracts, but was never ever meant galactically before Voyager.'

'Here Comes The Sun' certainly would have been an okay fit for the record. It's a pretty, delicate song, with a subtle yet satisfying melody and structure. It mentions a heavenly body, which *kind of* fits with the project, but the fact is, for me at least, it's no goosebump-raiser and it's nowhere near the best record The Beatles made. If The Beatles were going to be aboard Voyager, I for one would have wanted almost any other song to be their representative, 'Across the Universe' for example.

Ann was stoked anyway. Whatever the barrier was to 'Here Comes The Sun', its omission left the way clear for her favoured horse when it came to 20th-century popular music. Chuck Berry was definitely going.

While The Beatles missed that boat, another very British record with its own Abbey Road connection was being readied for embarkation. 'The Fairie Round' was written by 16th-century composer Anthony Holborne, performed and arranged by David Munrow.* Munrow was an influential performer, who in a very short career did a huge amount to further popular appreciation for Middle Age and Renaissance music. He led pioneering musical expeditions into the past, digging up, dusting off and recording a monumental multi-LP treasury of medieval music, with his touring and recording group of roughly 10 years, the Early Music Consort of London. By all accounts he was an astonishing human being, with vast reserves of energy and incredible skill – it is always boasted he could play 43 instruments, including the bassoon, the crumhorn and the recorder. In the UK he recorded more than 650 editions of *Pied Piper*, a programme full of 'tales and music for younger listeners' that enjoyed a five-year run on Radio 3. And he did all this before his death at the age of just 33.

'The Fairie Round', recorded at Abbey Road Studios in September 1973, was a perfect fit for the Voyager project as it shows off complex, wide-ranging musical form, while at the same time being very short – it's only one minute and 15 seconds long. Tim Ferris writes in *Murmurs* about how he felt it sat happily between various strands of the Golden Record playlist, forging connections between the horns and panpipes of New Guinea and the strings of Bach.

Finally, for all you Beatles fans out there still struggling with the disappointment, I would like to point out that side two of the Fab Four's second LP, *With The Beatles* (1963), kicks off with a cover of Chuck Berry's 'Roll Over Beethoven'. And who takes lead vocal? George Harrison. Isn't that all rather fitting?

★★★

---

* Munrow committed suicide almost exactly a year before the Voyager team had his work on the shortlist.

The Golden Record includes four pieces of purely, proudly American music. We know about Chuck, we know which Louis Armstrong track was on the ever-shortening list, we know they had a Native American piece in their sights, but there was another American track introduced to the team by Alan Lomax, and it's arguably the most haunting sound on the record.

The *Anthology of American Folk Music* was a six-album compilation released in 1952 by Folkways Records. It comprised 84 folk, blues and country music recordings that were originally issued from 1927 to 1932. It was put together by Harry Smith (largely from his own collection of 78s) and helped spark the American folk-music revival of the 1950s. It also led to many forgotten bluesmen enjoying a renaissance.

The compilation was divided by Smith into three two-disc volumes – Ballads, Social Music and Songs. The first included legendary recordings by the Carter Family, alongside Mississippi John Hurt and many others. The third volume had the likes of influential banjo player Dock Boggs, Blind Lemon Jefferson and that ever-smiling 'Dixie Dewdrop' Uncle Dave Macon. And sandwiched between them was Volume 2 (Social Music), which boasted a 1930 track by a performer known as Blind Willie Johnson.

Relatively little is known about the life of Blind Willie Johnson. He was born in January 1897 in Pendleton, a small town near Waco, to a sharecropper named George Johnson. His first instrument was a cigar-box guitar. He was not born blind. The story told by various biographers, and repeated by Tim Ferris in *Murmurs of Earth*, is that he was blinded by his stepmother when aged seven – she accidentally splashed him with lye water during a fight with his father.

He used a bottleneck-slide guitar technique that would influence generations after him, and he had a powerful, chesty vocal delivery. He was primarily an evangelist, performing religious songs on street corners, a brand of gospel blues, heavily influenced by his Baptist upbringing. His complete discography is about 30 tracks in length, all recorded during five different sessions in the late 1920s and early 1930s. His inclusion on the Harry Smith anthology helped revive interest in his work. And he was championed by folk mover-shaker Reverend Gary Davis, amongst others. One of his songs, 'Jesus Make Up My Dying Bed' (also known as 'In My Time of Dying'), was recorded on 3 December 1927 in Dallas. It would be covered by Bob Dylan in his self-titled debut in 1962 (the liner notes explain that Dylan used a borrowed lipstick holder as a makeshift slide). During the same session, Johnson recorded another track – a self-penned gospel blues number called 'Dark Was the Night', based on an old Scottish hymn and featuring penknife slide playing, it is thought, instead of his usual bottleneck.

The way Ann remembers it, Lomax played 'Dark Was the Night' to her and Tim on their first visit to his place in New York. If, back in 1977, you'd been asked to guess which blues tune might be included on Voyager, you might not have opted for Blind Willie Johnson. He wasn't completely obscure* but, compared to some of his contemporaries, who during the 1970s were gradually moving from legendary to mythic status – Robert Johnson or Lead Belly perhaps – he was an interesting choice. Rights issues aside, you might also have expected them to opt for a more obvious floor-filler from the likes of B.B. King or Johnny Lee Hooker. However, everything about 'Dark Was the Night' clicked. The title resonated with an object bound for the unimaginable blackness of interstellar space. That the rights were held by Folkways, who held the rights to many more songs in the shortlist, was a factor of convenience. And, alongside the song's spine-tingling atmosphere, the fact that it's essentially an instrumental, the only 'lyrics' being wordless moans, leaving nothing for an alien listener to interpret, helped it over the finish line. Aside from all that, it's utterly compelling.

---

* Fairport Convention included a song inspired by 'Dark Was the Night' called 'The Lord Is in this Place… How Dreadful Is this Place' on their second LP, *What We Did on Our Holidays*, in 1969. It's really good too.

This isn't the sound of a human being with the blues. This is the sound of longing, loss, despair. It's a 3a.m., death-comes-knocking song. And it's the perfect song to accompany two of the most remote, lonesome, human-made objects on a journey into eternal dark.

In 1945 Johnson was living at 1440 Forrest Street, in Beaumont, but only just. A fire had destroyed his home and, with nowhere else to go, he slept amid the ruins with his partner Angeline. The bed was open to the elements. He lay in soaking bedclothes, contracted a fever, and as the hospital wouldn't treat him, died at home in September 1945.

★★★

The music team were busy tracking down and listening to some of Robert Brown's suggestions. Armed with the 9 May letter to Carl, with its attached playlist and handy LP catalogue numbers, they had gone in search of the 'Pygmy Honey Gathering Song', chosen as a fine example of 'singing antiphonally'. The song in question came from the 1958 Ethnic Folkways record *Pygmies of the Ituri Forest*, recorded by anthropologist and ethnomusicologist Colin Turnbull,[*] who lived with and recorded the Mbuti in the Democratic Republic of Congo for many years. In the end, they found 'Pygmy Girls' Initiation Song' or 'Alima Song', a track also recorded by Turnbull which appeared on *Music of the Ituri Forest* (Ethnic Folkways Library, FW04483).

This is indeed an initiation song, performed during a ceremony to mark a girl's first menstruation. The Mbuti didn't tend to play instruments, they sang, sometimes in groups, each person in the circle given a single note, or in

---

[*] Turnbull wrote a bestselling account of his time with the Mbuti in 1961. *The Forest People* is still required reading for anthropology students, although – like many other works of the era – it is now seen as an idealised, romanticised account of a simple, harmonious forest life.

more conventional rounds and chants. Communal singing is part of the group's social glue. And the track in question is really good. You can certainly understand how its sophisticated yet unencumbered harmonies charmed the Voyager team. Its round structure means it feels like there's no beginning or end, just a stream of joyous consciousness that the listener is dipping into. It's an affirming sound, the sound of community, of celebration, of moving towards maturity and the creation of new life.

Brown also put forward a piece played on a shakuhachi – a Japanese bamboo flute. He suggested 'Shika no Tone', performed by Haruhiko Notomi and Tatsuya Araki, and included on UNESCO's *Musical Anthology of the Orient*. He wrote about how it would communicate 'varieties of tone colour'. In the end a shakuhachi piece called 'Sokaku-Reibo' ('Depicting the Cranes in Their Nest', although 'depicting' is often omitted from the title), in this case performed by Goro Yamaguchi, was chosen. It was recorded in New York City, circa 1967, and originally appeared on the wonderfully named LP *A Bell Ringing in the Empty Sky* (which *Rolling Stone* gave a ringing review in 1969).

'Depicting the Cranes in Their Nest' may put Roger Moore fans in mind of the creepy 'they're heading for the hill' guy from *Live and Let Die*. And if you fancy a laugh, I recommend popping onto NASA's YouTube channel to canvass opinion. One enlightened user commented: 'Honestly, what a waste. Five minutes of nonsense flute.' Another felt it needed 'more cowbell'. A third countered that it 'perfectly matches the mood of floating through deep space'.

Brown also proposed a piece he had recorded, an example of Javanese gamelan, a kind of orchestral arrangement of percussive instruments with pentatonic tuning, and choral and solo singing. The specific track is known as 'Kinds of Flowers'* and was performed by the Pura Paku Alaman Palace Orchestra in Yogyakarta, Indonesia, on 10 January 1971. You can predominantly hear bells and gongs, backed by loud

---

* Or as 'Ketawang: Puspåwårnå'.

human voices – according to Brown the different voices within the cacophony represent two of the nine sorts of flowers that represent states of Hindu philosophy.

Raga (sometimes raaga) is a word that comes from Indian classical music. According to most standard definitions, it is unique to Indian classical music (or at least has no direct correlation to Western classical traditions). If you imagine a band of musicians as a bunch of kids in a playground, a raga is like a jungle gym or climbing frame, over which they can climb, hang, scamper. It's a musical structure, on which they can improvise. And each group of notes, each raga, creates its own unique atmosphere or mood.

Alan Lomax didn't have much to say in terms of Indian pieces for Voyager. India and China were not his areas. However, Robert Brown had been very insistent from the beginning about an Indian raga named 'Jaat Kahan Ho', performed by Kesarbai Kerkar. As far as Brown was concerned, it was the beginning and end of the Indian question; it was the only record they need consider.

Kesarbai Kerkar, born in 1892 in North Goa, was a classical protégé, one of the best-known khayal singers of the latter half of the 20th century. She was particular about her work, however. So particular that, while she was celebrated as a live performer, her recorded output is fairly thin on the ground. You will often see her name written with the prefix 'Surashri', a title bestowed on her in 1948. It is literally translated as 'excellent voice'.

'Jaat Kahan Ho' was out of print. With Robert being so vociferous about it, the track had been on the shopping list for some time. For two weeks in May Ann had visited a number of record stores to track down a copy, but without success. By now, with time running short, she telephoned Brown, asking him for an alternative raga. He refused point blank. This wasn't one of his mere 'oh, this is quite a good one' recommendations. This, as far as Brown was concerned, was definitive – the finest example of Indian classical music ever committed to vinyl. That was all very well, but they had a flight to catch, and when Ann rang him they had three days left to complete the playlist. Ann pleaded with him, explaining that they might not find it, and if they didn't find it, Indian music would be missing from the record. But Brown held firm, urging her not to give up. The next day – with now just two days to go – Ann phoned him again. She had been working the phones, trying various other sources, contacting librarians, all to no avail. She promised Brown that she'd keep looking, but she begged him for a second option, *anything* that they could have in their back pocket. Again, Brown said there wasn't anything.

Eventually Ann's persistence paid off. She visited an Indian family-owned appliance store on Lexington Avenue, New York. There, in a brown box under a card table, she found three unopened copies of 'Jaat Kahan Ho'. She bought all three.

Go listen to it. Find a good copy and put it on. It's about three and a half minutes of hair-raising beauty. The mood is sombre, yet hopeful. The vocals yearning, yet powerful and defiant. It rules. Ann called Brown, thanking him repeatedly.

In the spring of 1977, around the time Ann was repeatedly thanking Brown, Kesarbai Kerkar had been retired from public singing for more than a decade. She died at the age of 85 on 16 September 1977, just 11 days after Voyager 1 left our atmosphere.

★★★

The remaining pictures were all about human beings – how we look, what we wear, the environments we occupy, the jobs we do, the things we make, the way we work and socialise. Drake would tell the local press around launch time how they chose images that were both informative and gave a 'balanced picture' of terrestrial civilisation, how they chose people from across the globe, rich and poor, and tried to include roughly equal numbers of men and women.[*]

---

[*]You can see this report in the 25 August 1977 edition of the *Cornell Chronicle*.

We have a gorgeous smiling man from Guatemala brandishing a machete (or at least a very large knife), a dancer from Bali, and a group of Andean girls in distinctive dress. There's a pair of craftsmen carving elephants in one picture, followed by a human riding an elephant shifting logs in another. There's an old man from Turkey, wearing glasses and smoking a cigarette. There's another old man walking through a field, with a small dog wandering alongside him.[*] There's a mountain climber on a needle of Alpine rock, a stroboscopic picture of a gymnast on a balance beam, a group of Olympic sprinters, and a school room.

Meanwhile every shot, before it could be stamped into the final sequence, had to be cleared with photographer and publisher. 'Most vividly, I recall calling Gaston Rebuffat in France,' says Wendy, 'whose photo was of a mountain climber on a narrow rock in the Alps. This was late in the timeline and we were desperate to get a hold of him.' Indeed, according to an interview with Drake in the *Cornell Chronicle* in August 1977, they finally located Rebuffat by phone somewhere in France at 2a.m. in the morning.

Many pioneering photographers made it aboard Voyager. Phillip Leonian's career spanned from the 1950s to the '80s, often appearing in titles such as *Look* and *Sports Illustrated*, and included the famous portrait of Muhammed Ali in a red-velvet robe and crown. Leonian specialised in capturing motion within a single still, and his stroboscopic photo of gymnast Cathy Rigby would become picture #71. He even sent Wendy another version from the same shoot, which she has kept to this day.

Next came the man–made world, starting with the Great Wall of China photographed by H. Edward Kim for *National Geographic*, its turreted back snaking up and along the brow of a distant hill, with human figures giving scale and perspective. This is followed by a UN photo showing a group of men in Africa building a wall from bricks, an Amish barn raising,

---

[*] This is one of three pictures (#44, #61 and #69) that show human beings accompanied by tame dogs.

then a simple hut, followed by a typical American house complete with white picket fence, and then a more modern-looking dwelling. This final domestic building was in fact the New Mexico home of British-American radio astronomer John V. Evans, taken by Frank. Next comes the Taj Mahal – chosen out of a long list of the world's most recognisable buildings and landmarks that included a Mayan pyramid and the Eiffel Tower – then city shots including Oxford, Boston, before the UN building by day and by night, and Sydney Opera House (still under construction).*

There are humans at work – a close-up picture of a man using a drill, a colour picture showing the inside of a factory (sent in colour to show the electric glow of the machinery), a woman using a microscope. Then come traffic and transportation: a street scene from Pakistan; rush hour in India; an Eckelmann original of Route 13 in Ithaca, with about four vehicles; another Ansel Adams picture, this time showing Golden Gate Bridge, with its spanning distance added above (1,280 metres); a plane taking off from Syracuse airport (another taken by Frank Drake); a train on the Boston–Washington line; an aerial view of Toronto International Airport, where Carl and Jon had their first meeting.

Finally the pictures return to astronomy and space: a radio telescope (specifically the Westerbork Interferometer in the north-eastern Netherlands, with cyclists in the foreground); the Arecibo Observatory with its 305m diameter noted on the image; a page from a book by Isaac Newton; the Titan Centaur launch (launching a Viking probe to Mars in 1975); an astronaut conducting a spacewalk; and the final transition to the music – a photograph of a quartet, a musical score and a violin.

Towards the end – image #114, to be precise – they included a second photograph by David Harvey. Harvey was born in San Francisco and raised in Virginia, discovering photography aged 11 after buying a used Leica with money

---

* Designed by Danish architect Jørn Utzon, the building was formally opened on 20 October 1973.

saved from his paper round. He was already a veteran of more than 40 photo essays for *National Geographic* by the time Voyager came knocking, and in just a few months' time he would be named 1978's Magazine Photographer of the Year by the National Press Photographers Association. According to Jon, Harvey's image #114 is one of the only shots chosen solely because it showed our planet looking rather nice. It was the last shot in colour, showing a reddening sky and a perfect silhouette of a line of 12 geese in flight with a few more on the water below. After this photograph, there were only two more images to come – the music transition – so this was a farewell image of Earth, the final shot aliens might see of our planet's surface and environment. When our planet boils to nothing in a few million years, this will be a final record of how a setting sun appeared to us.

These colour photographs caused problems as they took up three times the space of black-and-white images on the record, so they had to be worth it. The team only wanted to use colour images when it served the story of Earth they were telling.

The first colour image was that spectrum, followed by two of the Earth – one showing the entire planet (#12), the second showing the surface of Earth a little closer (#13). Picture #23 was recorded in colour too – a mid-1950s illustration of human anatomy taken from the *World Book Encyclopedia* rendered in colour so as to show the clear blues and reds that differentiated the veins and arteries. The next colour picture to appear in the sequence was the UN library image of a woman in a floral dress breastfeeding (#34), then came a beautiful picture taken by David Harvey of a man smiling up at a young naked child he's holding delicately in his arms (#35) – probably the most life-affirming shot in the whole collection, taken in Malaysia while he was there on assignment for *National Geographic* in 1976, and originally printed in the magazine in May 1977.

Image #36 is an interesting one, which I pick out at this stage as it illustrates the thought processes of the image team. It's not a bad photo, but it's not that interesting either. Unlike perhaps Stephen's flying insect, or the baby in its first seconds of life outside the womb, this image would not stop you in your tracks. Jon explains that, right at the start when they thought they would only have space for a mere handful of images, someone had suggested they have one image that showed a number of humans of different racial backgrounds doing something together. Three photographs that made the final cut all trace their inclusion back to this thought. One is the picture of Olympic sprinters, which includes a white man, two black men and an Asian man. Two others picked to illustrate race were taken at the UN International School in New York City: black-and-white image #74 shows a group of children gathered around a globe;* the other (#36) is full colour and shows a group of children, of various racial backgrounds, seated on the floor. Not only does it show a number of different types of earthling, but it also shows these beings seated in a number of different ways, showing more about the way our bodies work. There's a boy on his knees, a girl sat cross-legged and upright, a girl sitting with her legs splayed out to one side, and it shows hands held in various positions. Plus, as they're gathered in a circle, it shows us humans from all angles. So that's why a perfectly pleasant but otherwise forgettable image taken by Ruby Mera, a photographer working for UNICEF, made it aboard the Golden Record.

One of the last photographs in the transportation sequence shows a dramatic moment in Vivian Fuchs's Trans-Antarctic expedition of 1958. This overland crossing of Antarctica included Edmund Hillary in the team, who became the first person to reach the Pole since Amundsen in 1911 and Scott in 1912, and the first ever to do so using a motor vehicle – specifically snowcats. The expedition is also notable for the scientific data that was collected, including the first accurate

---

* The visible part of the globe is even showing the Middle East – the same part of the surface of Earth seen in photograph #13 earlier in the sequence. Genius!

CHAPTER EIGHT

Flowing Streams and Firecrackers

'Well, I've been meaning to talk to you about that, Carl...'
Ann Druyan

Ann had been given the assignment of finding a single piece of music to represent China. She telephoned Chinese-American composer Chou Wen-chung, who recommended a track called 'Flowing Streams' without hesitation. Coming in at around the seven-minute mark, it's one of the longest compositions on the record. It's also the oldest piece of music, being part of a longer work known as 'Towering Mountains and Flowing Streams', thought to have been composed some time between the eighth and fifth centuries BC.

'Flowing Streams' is played on a guqin, a seven-string bridgeless zither. The musician is Guan Pinghu (1897– 1967), who first learned the instrument from his father before becoming a teacher at the Central Conservatory of Music, Beijing. The guqin has a range of around four octaves, and there are three distinct sounds – 'scattered sounds', plucking open strings, 'floating sounds' created by string harmonics, and 'stopped'. With such a range of pluck, slide, strike and harmonic techniques, guqin notation is dizzyingly complex. There are in excess of 50 different techniques that must be mastered. Even the most commonly used are difficult to get right, and certain techniques vary from teacher to teacher and school to school. There are also a host of obsolete fingerings and notations, rarely used in modern tablature.

'Flowing Streams' left me a little cool on first listen. I thought it was atmospheric, interesting and certainly – to this Westerner, at least – representative of China's musical culture. It was clean, clear and, while not concise as such, free of dead wood – perfectly manicured. Tim writes about how it captures the Chinese philosophy of solo performance, of emphasis on single tones, as opposed to polyphony. On repeated listens, I do see why it caused a stir among the Voyager team, and why it jumped straight to the top of the heap marked 'China'. Played loud through good headphones or speakers, you can feel the performance. You can hear the player's fingers dancing over the strings. You can feel a subtle rhythm that shifts beneath your feet, like sand in surf.

When Ann called Chou Wen-chung he was at the Columbia School of Arts. He explained that – according to an article Ann wrote for the *New York Times* in 1977 – it would resonate with Chinese people, but that it must be a performance by the virtuoso. Ann found the Guan Pinghu recording, listened to it and knew it was perfect. Flushed with success, ecstatic that she had resolved the China question, she wanted to tell someone about it. Ann knew Carl was attending a conference in Tucson so she called his hotel. Finding him out, she left a message. Later in the day Carl returned her call, and they had a short conversation that completely changed the course of their lives.

It was Carl who made the first move, disclosing to her that he had returned to his room, found this message saying that 'Annie called', and asked himself why couldn't she have left him that message 10 years ago. She described this conversation on WNYC Radiolab's 'Space' episode in 2007, saying how her heart skipped a beat. This was Carl saying what had been unsaid between them for months but was becoming impossible to ignore – that he really liked her and wished they were together. Ann responded in a way that revealed she was in the same position as Carl, that she really liked him too. According to William Poundstone's Sagan biography, she replied: 'Well, I've been meaning to talk to you about that, Carl … Do you mean for keeps?' He responded that yes, he meant 'for keeps', he meant they should get married. She put the phone down and screamed. A few moments later the telephone rang again. It was Carl double-checking that what had just happened, *had* just happened. They were getting married? Ann confirmed that they were. She never mentioned the guqin.

Speaking to me in 2017, she said: 'Weeks would go by, and the four of us would go out together and then we got to work together and then we got to see each other more often. But throughout it wasn't until I found that piece of Chinese music on the telephone at a great distance – he's in Tucson, I'm in New York, Upper West Side – that even though we'd been alone together in a hotel room in New York … I mean, think of that. And we were smoking dope. We were both high and we still … neither of us knew … it was just out of the question. But on a telephone call … It was so overwhelmingly true. There was no avoiding it. It was a great truth. We both saw it and we both understood that it was right … It was like, "for keeps? Get married, right?" And it was "yeah, let's get married." Now, we had not kissed. We had not, you know … We had no reason to believe, especially in that time, that we should get married except for our instincts about each other based on these many conversations and wonderful evenings. But we were like: "We're going to get married." And all we have to do now is wait …'

Ann had previously confided in her friend Lynda Obst that she feared she was falling in love with Carl. And Poundstone describes that Carl invited Ann and Tim along to his friend and Harvard professor Lester Grinspoon's place during the Voyager project – and that, to Lester at least, it seemed obvious that Carl was interested in Ann.

Exactly when Carl fell in love with Ann is hard to say. The two obviously hit it off at Nora Ephron's party. Ann was smart, warm, with a life and upbringing that gave them plenty of common ground. The 'for keeps' phone call was a turning point that came out of the blue, and yet there had been signs that something was coming.

In Keay Davidson's Sagan biography he quotes from an interview with Gentry Lee (flight systems engineer at JPL and sci-fi writer). Lee recalled a dinner with Tim, Ann and Carl. Afterwards Sagan asked Lee if he thought Ann seemed attracted to Carl. And in the same book, Ann talks about one earlier conversation with Carl, that she realised afterwards was a sign of feelings that were to come, feelings that went beyond simple friendship.[*]

Carl met Ann at the Russian Tea Room, 57th Street New York. This was right after the Mars Viking landings – which would make it either late summer or autumn of 1976. Here Carl admitted that the conversations he'd had with Ann were the best conversations he'd had with a woman. There are a number of ways you could interpret that kind of statement. It could be called sexist – that he finds most conversation with women substandard or unstimulating. It could be seen as disclosure – an admission that he can find talking to members of the opposite sex awkward. But it's also a line-crossing, flirtatious thing to say. And if we believe that was a line crossed, what he said next left the line a couple of strides behind. Carl said he wished she was a man. He said it would make his life more simple. That sounds like a 'I like you quite a lot and you're messing with my head' kind of admission. Ann told Davidson that at the time she just deflected it, and it wasn't until much later that she truly realised it was a precursor to what would come in the 'for keeps' phone call.

Most humans might find themselves on one or other side of conversations like this at some point in their lives. A heart-quickening moment, when some ambiguous, half-said statement causes you to look up, to do a double-take, to go over the conversation in your head on the train home, only for the moment to pass, be shrugged off and lost in the blizzard of the everyday.

The 'for keeps' phone call, the love story of Druyan and Sagan, has become part of Golden Record lore, astronomy's

---

[*] See *Carl Sagan: A Life*, Davidson, pp.302–3.

June Carter and Johnny Cash. However, back in 1977 it wasn't simple. For a start there was still a lot of work to do. A great big emotional bomb like this could sink the whole project. They had to carry on as normal. Ann told Lynda Obst but, aside from that, she returned to her work. Carl, meanwhile, had just received word from NASA that the UN recording session was going ahead the following day. Someone from the project team had to be there to oversee. So Carl called Tim.

★★★

The day of the recording, 2 June 1977, was a shambles. Some committee members were there, others were not. The Cold War was still raging, so ... well ... um ... wouldn't it be quite bad if the Soviets weren't there? Yes, it would. Guess what? The Soviets weren't there. Ferris, who comes across as a relatively unflappable human, donned his project hat and tried to assess just how many languages were represented. There were only 15 nations in the room, and many of them spoke English. Brilliant.

If the disappointingly narrow spread of languages was a problem, a bigger problem was how much they all wanted to say. The whole idea of a record in space was a fairly strange one to communicate, but none of the delegates, it seemed, had been properly briefed. Tim explained, as best he could, how the record had very limited space and they needed to keep greetings to a minimum, but far from a few simple 'hellos', these were long rambling speeches, introducing both the speaker and their country of origin. Tim did his best to corral them into some kind of order, stressing that brevity was their friend, but the delegates were not to be dissuaded. Tim recorded the lot.

Writing post-launch, Sagan described how some of the greetings from this session turned out really rather well. He cites in particular the French and Swedish delegates, who both quoted pieces of poetry, the Australian delegate who thought outside the box and made his remarks in Esperanto, and the Nigerian delegate who spoke with unshakable confidence that aliens would know all about his home country.[*]

Tim and Carl listened back afterwards. What did they have? At first, it seemed like all they had was a whole load of verbose platitudes in a frustratingly narrow spread of languages, and only one female voice. Not only that, but they were now weighed down by political pressure to include the verbose platitudes. Politics meant that this tiny, finite metal disc, would now be forced to give up its valuable real estate to waffle.

In the immediate aftermath, with Tim and Carl still wrestling with the fetid cauldron of crapola the UN session seemed to have dumped in their laps, something else came up. The team had done a good job thus far of keeping the whole project under wraps, but now the United Nations had gone right ahead and blabbed to the press about what had happened. Not only that, they named Tim as a NASA official.

Suddenly everyone was unhappy. The record team didn't like that the secret was out. NASA didn't like that the secret was out. NASA also didn't like Tim going by NASA credentials he didn't have. Carl was annoyed about being told off – the record committee could not claim to represent NASA in any official capacity. And Tim was annoyed as he had never claimed to be a NASA official and now had a whole load of NASA officials annoyed with him for something he hadn't done. Bloody United Nations!

Just as Carl was dealing with all this flak and fallout, some unexpectedly good news came across his desk. Well, good news that later turned out to be bad news disguised as good news. Unbeknownst to Tim and the rest of the team on the ground, the Secretary General of the United Nations, Austrian diplomat Kurt Waldheim, had heard

---

[*] 'As you probably know,' he said, 'my country is situated on the west coast of the continent of Africa …'

about the record and written his own greeting. Not only that, he had recorded himself reading the message. And there were two further surprises: it was quite short *and* quite good.

The message ran like this: 'As the Secretary General of the United Nations, an organisation of 147 member states, who represent almost all of the human inhabitants of the planet Earth, I send greetings on behalf of the people of our planet. We step out of our solar system into the universe seeking only peace and friendship, to teach if we are called upon, to be taught if we are fortunate. We know full well that our planet and all its inhabitants are but a small part of this immense universe that surrounds us and it is with humility and hope that we take this step.'

See? It's good, isn't it? And perhaps more importantly than that, it clocked in at under 45 seconds. It was in. But it came with a cost: its inclusion triggered the most pointless addition to the whole project.

Whichever way you cut it, even with Kurt's eloquent passage, the UN greetings hadn't worked out as they had hoped. With judicious editing, and the help of some humpback whales, they would shape up all right by the time of final mastering, but at this point they had failed to represent the globe either in terms of language or gender.

Frutkin, the man who oiled wheels between Sagan, NASA, the State Department, and the UN's Outer Space Committee, next suggested a cocktail party. Why not get a whole load of Washington's ambassadors into a room, set up another mic and let 'em go. But Carl was now wary of politicians, especially drunk politicians. There wasn't the time for this – it was June already. So, still cogitating over the UN recordings, he looked closer to home: Cornell. They wanted to represent a global community in language and Cornell had a thriving population of linguists and foreign-language students on tap.

This is the part of the story that sees Linda Salzman Sagan take centre stage. She and Carl's assistant Shirley Arden did much of the leg work, sending out the call for help to the

wider Cornell community, and corralling them into action.[*]
Dr Steven Soter, who would later help co-write the series
Cosmos and its eventual follow-up, drew up a list of the world's
most widely spoken languages. Carl had given them the rough
target of recording the top 25. That was the minimum, with
the ambition to bag more if they could.

Shirley, Wendy, Linda and Steven began hitting the
phones. They started with Cornell language departments,
in turn calling individual speakers, booking them where
possible for one of the two recording sessions that had been
fixed on the calendar. By this time, of course, it was getting
towards the end of the semester, so the student and faculty
body was rapidly dwindling as people headed off for
summer vacations. Some languages were proving harder to
nail down than others. They called friends, and friends of
friends, asking for anyone who could speak certain
languages or dialects. A physicist named Bishun Khare –
who from 1968 to 1996 worked in Carl Sagan's Laboratory
for Planetary Studies – was 'almost single-handedly'
responsible for tracking down an array of Indian speakers
for the record.

The conduit through which humankind's languages were
to be preserved in our billion-year amber was a room at
Cornell. There Joe Leeming, a long-serving member of
Cornell's public relations department, set up the first recording
session. It took place on 8 June 1977 – six days after the UN
recordings. The second recording session took place on 13
June, this time helmed by David Gluck and Michael
Bronfenbrenner.

The candidates had been given no clear instructions. They
were just told they were greeting extraterrestrials and to keep
it short – Janet Sternberg, the Portuguese speaker, describes
the greetings as 'proto-tweets' in *The Farthest*. People arrived
in orderly fashion. Speakers were set up in the waiting room,

---

[*] Linda would also write the explanatory chapter, 'A Voyager's
Greetings' within *Murmurs of Earth*.

so the person waiting to record their greeting could hear the previous speaker's words. This, according to Linda, helped to create feelings of camaraderie and excitement as people waited their turn.

Frank's assistant Amahl Shakhashiri was drafted in to handle the Arabic greeting. 'We were told to be brief, and to speak clearly,' she says. 'The request came with no other instruction except to be brief. I could decide what to say. I thought about that and decided to say something that conveyed friendship and warmth, and a yearning for contact with the extraterrestrials. I wanted to convey that it was the yearning of our life to do that. So I chose my words carefully and rehearsed often – I wanted to make sure I represented the Arabic-speaking people eloquently.'

On the day of the recording she sat in the waiting room. She recalls feeling excited and nervous. She listened to the previous person recording their greeting – although she can't now remember which language that was – then it was her turn. 'I went in, sat down, and spoke my words. A few seconds, and it was over.' Her greeting was: 'Greetings to our friends in the stars. We wish that we will meet you someday.' She walked out, headed back to the NAIC offices in the Space Sciences Building. 'I remember feeling relieved, somewhat giddy and happy, and also wondering if anyone or any living creatures would ever hear it.'

The Japanese speaker, Mari Noda, is now a professor at Ohio State University. She was 25 at the time, had graduated from Cornell in 1975 and was working as an undergraduate teaching assistant.

'It seemed like a rather fantastic idea – so, so remote from our daily routine,' she says. 'The recording itself was quite simple. It was done in the basement language-lab studio, just like all the language-material recordings being done at the time. Later, when I heard some other greetings that my fellow recorders had made in other languages, I was astonished that they were quite lengthy. I was told to just say hello, so that's what I did. I clearly didn't comprehend what "greeting" meant back then. As a Japanese woman of the eligible age but without any prospect of marriage in sight, a strange thought did occur to me: what if some being from outer space "called", seeking lifetime partnership with me – or, by the time that would happen, more likely with one of my descendants – just because they were intrigued by my voice? How long would their "lifetime" be? Well, so far, no signs of such contact. I have a daughter who is about to start college, so I had better warn her.'

Even as the recording sessions were going on, they were still trying to find and recruit new speakers of languages not yet represented. People there to record one language often suggested a name of someone who might be able to speak another. This led to many people turning up very last minute.

An architecture student named Andrij Cehelsky voiced the Ukranian greeting. He was 23 and had just graduated from an intensive five-year architectural programme at Cornell. 'I stayed in Ithaca that summer to photograph my drawings and assemble an architectural portfolio, to unwind and just enjoy Ithaca. There was a small group of us architects doing this, it was pleasant and low-stress.'

Andrij (pronounced like Andre, but with a long e) was surprised when he received the call from Linda Sagan, asking if he would be interested in saying a few words in Ukrainian: 'I remember the phone going in my little kitchen – what an exciting moment! Though I had taken Sagan's "Introduction to Astronomy" course just before in the spring of 1977 and understood the gravity of the project, I did not see this coming.'

He agreed immediately. He called his mum in Rochester to talk it over, and contemplated a message that might include some verse by the great 19th-century Ukrainian poet Taras Shevchenko. However, as Linda had emphasised that it should be short, his final words included 'щастя, здоров'я і многая літа' ('wishing happiness/good fortune, health and many years') – a common phrase Ukrainians use when sending birthday wishes. He describes his recording session the following day:

'I registered, signed a sworn statement – signing off to NASA all rights of the recording, submitted a hand-printed copy and translation of my greeting. When my turn came, I went into the recording room, sat in front of a microphone and recorded the greeting. I cannot recall if we did multiple takes. That was it! I walked back to my apartment, slowly comprehending what had just transpired.

'Though born in the US, I grew up speaking Ukrainian as well as English. Ukraine had been absorbed into the Soviet Union after the Second World War and became a captive nation. No free press, no free representation in the outside world. The Ukrainian language was deliberately marginalised while Russian was promoted. There were no Ukrainian foreign students at Cornell, and Ukraine's seat in the UN was just another vote for the Soviets. So the Ukrainian diaspora worked hard to preserve the language, culture and historical memory. In that context, besides representing our planet, I was also representing Ukrainians who did not have a free voice in 1977.'

The Cornell greetings are utterly charming. In the end the team far exceeded the 25-language target, with a rapid-fire, smash-and-grab operation that bagged 55. In *Murmurs* you can see the results as a table showing the language, the name of the speaker, the comments in that language, an English translation, then a rough approximation of the number of human beings on the planet who spoke that language.

There were some disappointments, however. There were no-shows and gaps that the team just ran out of time to fill – there is, for example, no Swahili greeting – but what they did get was an unscripted array of idiosyncratic warmth. The Indonesian speaker, Ilyas Harun, says: 'Goodnight, ladies and gentlemen. Goodbye and see you next time.' Jatinder N. Paul, speaking Punjabi, says: 'Welcome home. It is a pleasure to receive you.' Stella Fessler, speaking in Cantonese says: 'Hi. How are you? Wish you peace, health and happiness.' The Mandarin greeting, spoken by Liang Ku, is a personal favourite: 'Hope everyone's well. We are thinking about you all. Please come here to visit when you have time.' The Amoy (a Min dialect) speaker asks whether the 'friends of space' have eaten yet. Radhekant Dave, speaking Gujarati a dialect of western India, sounds positively desperate with 'Greetings from a human being of the Earth. Please contact.' And Gunnel Almgren Schaar says in Swedish: 'Greetings from a computer programmer in the little university town of Ithaca on the planet Earth.'* They even managed some extinct languages, including Latin and ancient Sumerian.

Pure chance had selected this band of ordinary souls from the Cornell campus for immortality. Many were studying students, others faculty members. Professor of Classics and comparative literature Frederick M. Ahl recorded the Greek, Latin and Welsh greetings. David I. Owen, professor of ancient Near Eastern and Judaic studies, voiced greetings in ancient Sumerian, Akkadian, Hebrew, Hittite and Aramaic. Maria Nowakowska Stycos, the wife of another Cornell professor, was studying for her PhD and provided the Polish greeting. The Bengali speaker Subrata Mukherjee is now a professor in the mechanical and aerospace engineering department at Cornell. The Rajasthani speaker Mool C. Gupta, a postdoctoral fellow at Cornell, who said 'we are happy here and you be happy there', is now a professor at the University of Virginia. Swedish speaker Gunnel Almgren Schaar went on to become a professor in the computer science department at Grambling State University.

Before the greetings tapes could be signed off for mixing and mastering, there was one more ingredient. They wanted a child's voice on the record, to greet the cosmos on behalf of youngsters everywhere. For this final assignment, greetings team leader Linda simply went home, where she and Carl recorded the very sweet-sounding six-year-old Nick Sagan

---

* You can listen to the greetings via NASA's Soundcloud page: soundcloud.com/nasa/sets/golden-record-greetings-to-the.

saying: 'Hello from the children of planet Earth.' Nick* recalls being plopped down in front of a microphone, and being asked what he'd like to say, if he could say anything, to extraterrestrials. In *The Farthest* he recalls watching the VU meter needle as he spoke into the mic. In interviews he describes how all this didn't seem too weird back then – his father was Carl Sagan, after all – but how today Voyager leaves him feeling wistful, that some tiny fragment of his childhood is preserved forever, a moment in time, moving further and further away.

Many of the Voyager greetings veterans describe similar feelings – a gradual, dawning realisation that these tiny fleeting moments, memorable but transient, were now forever enshrined in these distant artefacts. Andrij says: 'I feel very special to be part of this exclusive club of 55-plus speakers on the Golden Record. A relatively short life experience has put me in the history books, and in a relative sense, a trace of my being has been immortalised. The experience has also solidified my interest in astronomy and I feel part of the family of space travellers. And just as in 1977, I understand how important it was to represent Ukrainians at such a difficult time in their history.'

Amahl adds: 'It wasn't until later that the enormity of it hit me in an emotional sense, after the work was done and there was finally time to sit and relax and contemplate. Today, I am 40 years older than I was then, and though I am the same person, I am not the same. And those years and life experiences have deepened my appreciation of the value of the work we all did to make this Golden Record. It was the gig of a lifetime and I am truly honoured and humbled by it all.'

★★★

---

* Today Nick is a writer. He wrote the *Idlewild* series of novels, episodes of *Star Trek*, produced a two-part *Alien Encounters* TV special with SETI, and helped promote the recent *Cosmos* reboot.

The picture team wanted to show human anatomy in detail. And they seemed to have found the very thing – a set of acetate overlays from the *World Book Encyclopedia*, each taking a section of the human anatomy. But there was a problem. While the eight illustrations were essentially perfect in all respects, showing human surface tissue, skeletal structure, circulatory system and musculature, they were covered with lots and lots of tiny black numbers. These tiny black numbers related to a key printed in the original book in which all the different parts of the body were named. However, this key wasn't going with Voyager, and even if it had no one would be able to read it, so there was no point in these tiny black numbers. They served no purpose. Jon got in touch with the World Book team. It seemed there would be no problem providing a set of the original acetates, but they did not have any copies of the illustrations *without* the tiny black numbers. And so Linda set to work, spending hour upon painstaking hour, carefully painting over the numbers, matching the colours behind so that the numbers would be hidden.

It seems appropriate that the woman whose drawing of a vulva was obliterated by NASA should now be tasked with covering up some unwanted black lines on their behalf. However, not only did it take her ages, it also didn't work. Following hours of hard graft, the paint dried and flaked off. The black spots and numbers returned, and by now there was no time to do anything about them. Soon the images had to be whisked away for their final conversion to audio ahead of mastering. And so the anatomy sequence, though perfect in many respects, is sealed aboard with hundreds of tiny black dots and symbols whose meaning is likely to cause alien readers endless confusion.

All the anatomy diagrams, which form the basis of images #18 to #25, were neuter. There was just a void where the sexual organs should be. By this point, Jon was a man possessed, a man completely immersed in the imagined headspace of an alien. There were more images to come that would tell the story of human reproduction but, the way Jon saw it, unless they added something here, the jigsaw would be missing a vital piece. So, to the asexual image #25 in the anatomy sequence, they added the male and female symbols that crop up elsewhere in the sequence (specifically image #32), and diagrams of the male and female reproductive organs.

Finally, there was image #26, taken from *Life: Cells, Organisms, Populations*, a full-on sex-education-style cross-section diagram of both male and female reproductive organs. To this, again, the team added the female and male symbols, and a scale in centimetres. The anatomy sequence was ready. The images would take their place right after the DNA diagrams, and just before the pictures of egg, sperm and foetus.

Now all they had to do was sneak all these genitals past NASA.

★★★

On Friday 3 June 1977 Ann Druyan walked into the New York University Medical Center. She was there to record one of the final pieces for the sound essay: Ann's EEG patterns. They already had a heartbeat, footsteps, laughter and speech, so why not go all the way and record thoughts and brainwaves – specifically the voltage fluctuations resulting from ionic current within the neurons of Ann's brain? The more 'far out' aspect to this plan was the thought that some advanced alien race might one day be able to interpret and even reconstruct her thoughts. Ann met Dr Julius Korein, and with Tim's help they fixed an audiotape recorder to the medical data recorder, and Ann was left alone in a room to meditate for an hour.

Recalling this experience years later, she described how these moments were the fulcrum around which her life revolved. It was two days after the phone call with Carl. Ann knew there were clouds gathering, but for now things had to carry on as normal. A scandal could be like a wrecking ball, derailing the project just as it seemed to be coming together. The Golden Record came first. And while the Nora Ephron party had been the start of her relationship with Carl Sagan, this hour in a New York medical centre, attached to an audiotape recorder, was a powerful moment of calm.

Ann had prepared for the recording. She didn't intend to just lie there and read a magazine. This was for a record going into space, so she had come along with a mental checklist of specific things and subjects she would think about – ideas, individuals, moments in history. Writing about it the following year, in *Murmurs of Earth*, she mentions how a couple of 'irrepressible' facts from her own life were also in her thoughts. Years later she would expand on this: she was in the full-blown throws of recently realised love.

It's about the most human situation there is. Most of us know what it's like to yearn for a human being, to want to be alongside that human being for every waking moment. The difference is, of course, that when we experienced it we might have been trying to get to sleep in a dormitory in Sussex, or returning from a tour of duty in Afghanistan, or having just taken a promotion, or drinking coffees in a canteen, or having just danced drunkenly with someone at a party. Indeed, for many of us, these kinds of thoughts and epiphanies assault us right around the same time the hangover kicks in. For Ann, all these thoughts and emotions came over her while having her brainwaves measured by a machine ahead of having the audio put on two metal records that would drift in space for a billion years. The Voyager Golden Records have the sound of brainwaves from a human being falling in love with another human being, but not yet *with* that human being.

Eventually, when it came to the final 'Sounds of Earth' mix,[*] the hour of recorded thought impulses would be

---

[*] Ann Druyan's brainwaves form the penultimate piece of the sound essay. The final segment is a recording of a pulsar made by Frank and Amahl, specifically the warbling radio signal projected by the spinning star CP1133.

compressed into about a minute. Sped up to such a degree, it sounds like crackling static, or firecrackers. Ann said in a radio interview: 'I was thinking in this meditation about the wonder of love, and of being in love ... Now, whenever I'm down, I'm thinking, still they move, 35,000 miles an hour, leaving our solar system for the great wide-open sea of interstellar space.'

CHAPTER NINE

Mixing and Mastering

'Any record to go through a studio of the level of Columbia Records New York City in the 1970s is going to sound good. Those were the best engineers in the world.'

Tim Ferris

The business end of the Voyager Golden Record project came to its hectic conclusion in New York City. The picture team were still working feverishly in Ithaca, as Tim and Ann descended on CBS Columbia at 49 East 52nd Street, Manhattan. This was to be their home for the best part of two weeks as they oversaw the mixing, mastering, sequencing, clearing and cutting. Tim says: 'I really had to work pretty long hours to make sure we got it done on schedule because there were an awful lot of moving parts.' Those moving parts included all the sequencing and mixing of the sound essay, the greetings, the music and, let's not forget, the audio-converted pictures that hadn't even been delivered yet.

For the sound essay alone, Tim and Ann had returned to New York with around 50 sounds. They had armfuls of recordings in all sorts of formats, supplied on studio master tapes, on quarter-inch reels, and on original LPs. There was borrowed vinyl from Alan Lomax, shop-bought LPs, tapes from archives, yet more recordings from the Library of Congress. They had tapes of UN recordings and whale sounds, and a further selection of Cornell greetings was en route. Thanks to Mickey Kapp putting the Electra Sound Archives at their disposal (and even hand-delivering some individual tracks), they had material on tap for the sound essay, but they still had to sequence and mix the thing – which was no easy task – and they were still missing a few vital ingredients.

For all technical wizardry, CBS put Russ Payne in charge.
Russ was, by all accounts, a brilliant recording engineer. By
1977 he was one of the old guard, an experienced hand who
favoured sharp suits, thin ties and a neat side parting. He
had a narrow, angular face, a bullet dimple on his chin, two
sets of laughter lines and fine crow's feet beneath intense
eyes. He would helm the project and was present throughout
the entire mixing and mastering process. Writing about
Russ the following year, Ann describes him warmly as a
calm, cowboy-accented character who would eat fruit and
smoke cigarettes. And when it came to the essay at least,
most of the engineering was carried out by Russ and Tim
working behind the industry-standard hardware of a
16-track Ampex.

There were plenty of other CBS bods who would lend a
hand. One of them was a Brooklyn sound guy named Jimmy
Iovine. Tim had been in touch with John Lennon about the
project back when The Beatles were still in the running.
While their music was not destined to make the final cut,
Lennon would still impact the record by recommending
they use his engineer, Jimmy. Jimmy was still near the start
of his career, but he had already worked with Lennon[*] and
on Springsteen's *Born to Run*. He'd also contributed to
Meatloaf's *Bat Out of Hell*, due for launch around the same
time as the Voyagers. Alongside a pretty jaw-dropping roll
call of hit albums, Jimmy would work in the music
departments of a couple of good John Hughes films in the
1980s (*Sixteen Candles, Weird Science*), start Interscope
Records in the 1990s (Eminem, U2), and found Beats
Electronics with Dr Dre in the 2000s (um, headphones). For
now he was an engineer. Tim told me: 'Jimmy was there for
inspiration and ideas; he helped us physically mix the Sounds
sequence. When not needed, he hung out at Bruce

---

[*] Jimmy had been a studio assistant at the famous Record Plant in
NY. When he showed up to work on Easter Sunday, his bosses were
so impressed they let him work with Lennon.

Springsteen's recording session down the hall, which was
fine with me as I knew he was building a career and could
use the contacts.'

There were plenty of sounds that still hadn't been found.
They wanted to include, for example, a kiss. This was again
one of those sound effects that you'd think would be easy to
find but had proved tricky. The problem was they wanted to
record something genuine – but not too 'smacky'. In
Murmurs Ann describes how Jimmy, in the studio that day,
offered his forearm – he was sure he could create a believable
kiss by sucking it. But Ann wanted something real. In the
end, the sound on the record is Tim kissing Ann on the
cheek.*

Tim worked more or less full time at the studio. When
working on the Ozma Records reissue in 2016, he described
how the 'Sounds of Earth' required several people
simultaneously to have their hands on the mixing-board
sliders as he conducted. Ann was there a lot of the time too
but was also still making frequent field trips to track down
missing pieces of the jigsaw – the aforementioned Indian raga
hunt, for example. Carl would also pop by from time to time,
to charm the CBS team and listen to how things were
progressing.

★★★

When I was young, my father attempted to explain to me
how satellites worked. He had a piece of paper and a pen. He
drew a picture of the Earth, a large circle, with a stick boy
standing on top. Then, like a million science teachers before

---

* In the final mix of the sound essay, the kiss is followed by 'Mother
and child' – specifically the very first cries of a newborn infant, and
then a six-month-old baby being soothed by its mother. This audio
was provided by Dr Margaret Bullowa at MIT and her then-assistant
Lise Menn, who is now a linguist and professor at the University of
Colorado.

him, he asked me to imagine throwing a ball. A line indicated the path of the ball. The first throw went a short way along before curving down towards the circle of Earth's crust. Then he asked me to imagine throwing it harder. The second throw went further, coming to rest a little further around the circle. The third went further still, resisting gravity horizontally, before falling back to earth. Now he asked me to imagine throwing it *even* harder, so hard that it never reached the surface but kept on falling round and round in orbit.

The picture team, like the record team, were now working towards finalising and recording the pictures. This needed to be done ASAP, as they knew that Tim was by now pulling things together at CBS in New York and so he needed that tape. But they wanted to make sure there was a picture of a book. It seemed an important human trait to attempt to communicate – that we write, record and share information in these printed book thingies.

Jon Lomberg visited Donald Eddy, the curator of rare books at the Cornell University Library. Here there were all sorts of scarce and valuable volumes, including a Shakespeare first folio, but Jon was interested in a particular book suggested by Philip Morrison. Specifically he wanted the 1728 English edition of the third volume in Isaac Newton's *Philosophiæ Naturalis Principia Mathematica*, and it was the diagram on page six of book three, *De mundi systemate* (*On the System of the World*), that most interested him. Just like the picture my father drew for me when I was young, the diagram in question shows Earth with an imagined cannonball being fired, faster and harder, until it achieves escape velocity. Jon says: 'Philip Morrison suggested the page because it was the first time in history that we had shown how you could launch something in orbit.'

As usual, Jon wanted to cram in as much information as he could. So this photograph, one of the very last to appear, doesn't just show the page in the book, it shows a finger and thumb, pulling back the corner of the right-hand page, as if about to be turned, revealing the mechanics of printed books.

He says: 'I have a particular fondness for that image for a number of reasons, not the least of which is that it is my thumb in the picture. I wish I'd cleaned my fingernails!' The picture sequence really was going out on a high.

Amahl was kind enough to share with me some pictures, showing their workroom at Cornell from these final stages of the project. There's Amahl leaning over some slides. There's a camera. There's the desk covered in coffee-table books with bookmarks and notes. There's the projector. There's a video camera, filming another also-ran version of the licking, biting, drinking photograph, and there's that same photograph being televised on a small black-and-white screen.

The fact that the images and sound hadn't been made to work more closely together, so that sound could correspond with each photograph, was a lasting regret. Frank voiced this regret in his 'Foundations of the Voyager Record' essay in 1978, and voiced it again to me on the telephone 40 years later. But while this was no time for perfectionism, even as the image sequence drew to a close, there was one final chance to realise this thwarted ambition.

The pictures had been arranged broadly chronologically, just like the sound essay, telling an evolutionary story that climaxed with humankind's ultimate technological leap – to escape our atmosphere and journey into space – with two NASA pictures in the form of an astronaut in space, and another showing the 1975 Titan Centaur Launch at Cape Canaveral. But how should the picture sequence end? Jon's thoughts went to the alien audience and concluded that the images should end by teeing up the music that was to follow on the Golden Record.

'It seemed to me that if I put myself in the role of extraterrestrial, which is what I viewed my job description as being, then the music might be some of the most mysterious parts of the whole record ... You know a sound of thunder? Well, that's easy to recognise, there'll be thunder on other planets. Same with rain and surf. Sounds made by air passing over soft tissue ... those kind of sounds, they're possible to interpret. But what is music – to an alien? Is it earthling speech? Is it the call of an animal? What is it? So I thought there needed to be some explicit way of trying to explain what music is. That is the only example where we did that and that was my idea. Well, we have these sequences of Earth, how do we end it? What's the final picture? And I thought, "well, what comes right after the pictures – the music…"'

The final two images in the picture sequence are 'String Quartet' (#115) and 'Score of Quartet and Violin' (#116). The quartet in question is in fact the Quartetto Italiano, founded in 1945 by Paolo Borciani (1st violin), Elisa Pegreffi (2nd violin), Lionello Forzanti (viola, replaced in 1947 by Piero Farulli) and Franco Rossi (cello). Although this quartet did record a version of the final piece of the Voyager playlist – the Cavatina from Beethoven's *String Quartet No. 13* – it was not the Quartetto Italiano's recording that made the Voyager record. However, the photograph of them served other purposes. It pleased Jon because it clearly showed a number of human figures playing similarly shaped stringed instruments of different sizes – violin, viola and cello. It showed the mechanics of playing notes, of bows being drawn over strings. It also showed that the making of music was often a social activity – that groups of human beings came together to create music.

'One of the sounds that should be easy to decipher is the vibration of a string. That will be a universal. A string being plucked or bowed on any planet in the universe will make the same overtone series. Doesn't matter what the atmosphere is, it doesn't matter what gravity is, it doesn't matter what the sun is, the overtone series of a vibrating string is going to be the same.'

The final image (#116) is a combined one: a lone violin on its side against a black background, directly above a page of musical notation from the Cavatina from Beethoven's *String Quartet No. 13*. Immediately after this last image is encoded on the record are a few seconds of the sound of that piece of music. So the aliens will receive a picture of humans playing instruments, the instrument itself, some musical notation, and a few seconds of those instruments playing that piece of music, all in a direct sequence.

★★★

Thanks to delegation, Carl had been able to keep up with his day job during these busy days in May. By now, however, he was focusing solely on the Golden Record, and it was time for him to earn his crust. It was all rather worrying. All this time, all this work, and everyone knew that NASA could can the whole thing and there would be nothing the team could do about it. It was now early June. The mixing of the greetings, the music and the sound essay was coming together at CBS.

In *Murmurs* there's a description of NASA officials actually visiting CBS Records itself to check the contents. Carl was there. You might expect high emotion, confusion or angry rejections. This was art rubbing up against science, bureaucracy rubbing up against creativity. It seemed to Carl, anyway, that they were only interested in ticking some boxes, making sure nothing too racy or potentially embarrassing was trying to sneak aboard. Carl describes the officials' muted responses – from recognition of 'Johnny B. Goode' to 'bland approval'. Indeed the only fallout came the next day, when Carl received an agitated NASA call from some administrator who was concerned there wasn't an Irish tune on the record. Carl pointed out there was no Italian opera or Jewish folk music either.

However, Tim doesn't remember it quite like this. He remembers Carl having one critical meeting, in Washington DC, after which he remembers Carl reporting that they weren't really interested in hearing the music at all – they were much more concerned with the images. Certainly this Washington DC meeting took place. And, for another perspective, we can revisit Jon's manuscript. Let's imagine him and Frank like a pair of nervous uncles pacing a hospital ward, waiting for news of a birth.

'After we had selected all the images for the picture sequence and record cover, Carl took everything to Washington for a review by NASA Headquarters,' he writes. 'Frank and I waited nervously back in the Space Science Building at Cornell. Carl had promised to phone us from the meeting to tell us if NASA had any problems with the picture sequence. Frank and I were both at the thin edge of exhaustion, too many keyed up days and sleepless nights.'

When Carl did finally call, Frank put him on speakerphone. Carl was apparently surrounded by NASA bureaucrats and lawyers. Jon writes: 'He sounded as nervous and tense as I ever heard him, though he was trying to mask it.'

'Hi, Carl,' said Frank. 'How's it going?'

'Well…' said Carl, 'they like *almost* everything. The only picture they have a problem with is the nudes.* We must remove them, or cover up the genitals.'

Jon immediately flared up.

'That would defeat the whole reason for including it!' he fumed. 'And dropping it will reduce the clarity of the whole reproduction sequence.'

Frank silenced Jon with a gesture.

'Anything else?' he asked calmly.

'Well, they think the man looks too much like a surfer. Can we make his hair less wavy?'

This was too much for Jon who sat down in disgust.

'Is that all?' Frank asked again.

'Yes,' said Carl. 'Otherwise everything is okay.'

And the phone call ended there.

Jon writes: 'I was not pleased. Strung out as I was, I had no tolerance for NASA's caution.'

---

* The photograph in question came from a book called *The Classic Nude* by George Hester. It shows a pregnant woman holding hands with a man. They are both naked.

Meanwhile, Frank was positively beaming.

'Don't you get it? They approved our message. *It's going.*'

They walked back to their workroom. They removed the photograph of the nude pregnant couple from the sequence of slides spread out on the light table. Frank picked up the corresponding silhouette slide, with the figures shown in black outline, the foetus visible in the woman's stomach.

'How can we change it?'

There was now only an hour or two left until the sequences had to go to Colorado for recording. There was no time to create and photograph another silhouette. Frank said they would have to change it directly on the slide. Frank handed Jon a fine-point black marker and a scalpel saying 'Just do your best.'

At first Jon was too angry. He refused, pouting: 'It's bullshit what they want,' he said. 'And I can't do it with these tools.'

Frank sighed, picked up a magnifier and began to go about trying to retouch the slide himself. Jon relented, took the pen, the knife and the magnifier and sat down to 'scratch and dot the man's face into ethnic neutrality'.

A few months later Jon received a letter from a NASA lawyer explaining their decision and countering his contention that they had chosen a photo that was explicitly unerotic. 'Some people find naked pregnant women extremely erotic,' it read. 'I would suppose this to be among the most unusual phrases ever to appear in official NASA correspondence.

'In the next two decades I saw both how bureaucratic NASA was and also how bureaucratic they had to be,' Jon writes. 'Given the craziness that passes for sanity in Washington, and seeing various tempests over art funding, exhibits at NASA, and the indiscretions of politicians, I guess it was too much to expect NASA to take the long view of interstellar messages. The only audience NASA could afford to care about was the audience on Capitol Hill.'

Given NASA's timidity about nude humans, it does seem surprising that they did not also seek to eliminate or at least paper over parts of the anatomy sequence, or the naked bottom in image #62, or the picture of the nursing mother. Perhaps she got by without comment because the nipple was not exposed. Thomas J. Prendergast, picture librarian at the United Nations, also wrote to Jon after launch: 'You may be interested to know that the appearance of the photograph of the nursing mother in Malaysia accompanying some newspaper stories about Voyager, prompted over a thousand requests for copies of that picture. Most of them came from members of La Leche League, which promotes breast feeding...'

Carl visited Tim in New York later that evening. Tim says 'I remember he flew down from Ithaca to Washington. The flight was diverted by bad weather to, like, Baltimore. They took a train or something, got to NASA, had the meeting. Got on a plane. That one was diverted too, also to Baltimore again ... By the time he got to my place it was 11 o'clock at night, he was literally in tears. But he was cheered that, so far as the music goes, they just don't care. They didn't even really want to hear it. All they cared about was the nude photo. I think he kind of bulkheaded off NASA from our team and that was probably a good idea. We didn't see them they didn't see us till after it was done. And I think that was a smart way to do it.'

★★★

With the last pieces in place – the Newton diagram, spacewalk, rocket launch, sunset, string quartet and Cavatina score – and the entire sequence cleared by NASA, it was time to convert the slides to audio.

Val had found Colorado Video, that firm based in Boulder, Colorado, who could help them with this never-before-needed tech. The firm offered its equipment, expertise, personnel and time entirely as a public service. Frank describes in *Murmurs* how the president, Glen Southworth, personally assisted into the 'wee hours' recording the pictures. Technicians and engineers spent hours operating the equipment, making adjustments, tweaking dials and knobs to optimise quality. They would also give up time so the nervous picture team could have a complete trial run prior to the final recording session. Not only that, but staff at Colorado Video helped negotiate the loan of a Honeywell 5600C recorder – Honeywell Inc was based nearby in Denver. This was a portable 14-track reel-to-reel tape recorder that, in terms of fidelity, was the best machine for this odd job. It was advertised as 'portable' but it still had some serious heft. It was light blue in colour, about the size of a large, fatter-than-usual briefcase.

Unlike that working set-up at Cornell, the Colorado technicians did not project the photographic slides onto a piece of paper taped to the wall. The images were brought to and scanned at Colorado Video using a Dage Video Camera and the all-important 321 Analyzer, the output of the 321 feeding to the Honeywell recorder.

Judd Johnson, who was 33 at the time, was the fourth employee at Colorado Video. The company built all kinds of specialised video processing equipment for various applications, including some designed for astronomer Dick Dunn at the National Solar Observatory at Sunspot New Mexico. When Judd joined in 1970 the firm was located in 250 Pearl Street, West Boulder. 'Everything was low budget,' he says. 'We had built our own workbenches, wired the place, cleaned up after ourselves.' It sounds like it was a fun, ad hoc sort of place to work, with everyone mucking in, helping out on different tasks and types of work. During downtime, they created an indoor archery range. 'Because we were shooting down the aisles between the workbenches,' says Judd, 'we could only practise after hours or while everyone was on lunch break.'

By the time of the Voyager project the company had grown to around 14 employees. 'For as long as I knew him, Glen Southworth lived and breathed Video,' says Judd. 'He had conceived of a technique using the synchronising pulses common to the video format of the era, to derive the intensity value of the video signal at any point in the image/raster … He developed his first instrument, which ultimately became the Model 321 Video Analyzer.'

It was this Analyzer that enabled the conversion from video to audio. Another employee, Wyndham Hannaway, carried out the actual encoding. He was 27 at the time, working as an entry-level technician, which involved hand-wiring, troubleshooting, manufacturing products, presenting at trade shows, travelling to clients and some special projects. Wyndham didn't have any first-hand contact with the Voyager picture team – all of that was handled by Glen – but he does remember the work. Indeed, he and Judd are the only two Voyager veterans from the firm who are still alive.

The Golden Record encoding was carried out on a project bench that Wyndham hand-rigged in a darkened area of the building. The slides were placed on a light box, a black cardboard mask was made to cover the light box apart from a cut-out to hold the slide. A vertical copy stand held the standard-definition silicon-target vidicon tube camera built by Dage, which had a Nikon 55mm macro lens attached. A Kodak filter holder was also attached to the lens so that Wyndham could manually swap the red, green and blue filters for the three-colour exposures. Each video frame was 'frozen' in a disc-based video framestore, then the Honeywell was run for a minute to capture the slow-scan signal.

Through the various road tests of the system, the final black-and-white images took up around six to eight seconds of audio – much less time than Frank had guessed back in Honolulu.

Wyndham says: 'The one-inch tape from the Honeywell DC recorder was shipped to the Voyager team, and I drove the recorder back to south Denver to the Honeywell plant where we had borrowed it.' Phew!

★★★

It looked like those UN recordings might work out all right too. They were being saved in the editing suite at CBS. While there was political pressure to include everyone who had contributed, there was nothing in the small print about including *all* of each message. And the heavily pruned-down mix does include some lovely stuff. Bernadette Lefort quotes from Baudelaire's 'Les Fleurs du Mal'; Anders Thunboig of Sweden quotes the 1945 poem 'Visit to the Observatory' by Harry Martinson; speaking Persian, Bahram Moghtaderi sent a greeting from the people and government of Iran; there are examples of German, Indonesian, Efik and Flemish; Syed Azmat Hassan of Pakistan speaks in Punjabi directly and disarmingly to some imagined 'friends' in space;* Samuel Ramsey Nicol of Sierra Leone wishes the cosmos good luck; and the US's own James F. Leonard extended his 'greetings and friendly wishes to all who may encounter this Voyager and receive this message'. Finally the Australian member, Ralph Harry, recorded his message in Esperanto.†

Then, in the studio, came Tim's masterstroke: to mix the cut-back UN greetings with whale song. They had all agreed with Ann – it was a lovely idea to not only include human greetings, but also include some greetings or sounds from a species that shared our planet. So in the mix, behind the rather formal voices of our UN delegates, come the joyful sounds of some whales. We can have no idea what they're saying, but it is pretty clear they're having more fun saying it than the UN Outer Space Committee.

When writing about his reaction on first hearing Kurt Waldheim's unexpectedly concise message to the cosmos, Carl used words like 'sensitive', 'graceful', 'appropriate'. It had to go in as far as he was concerned. But now that a greeting from the head of the UN was a shoo-in, it suddenly seemed not only right but necessary that they should also give the

---

* 'On behalf of my countrymen I am sending a message of friendship and greetings to our friends in space …'
† His excerpt, translated, reads: 'We seek to live in peace with the people of the whole world, of the whole universe…'

President of the United States the chance to contribute. Uh-oh.

With the dread of what energy and time-sapping bureaucratic cogs the White House might sling in his path, Carl put in a call to Dr Frank Press, President Carter's science adviser. He also contacted the top NASA administrator. NASA was pro: anything that brought positive attention on the project from the American government was fine by them.

There was another wait until, a few days later, they had word. President Carter *did* want to contribute, but instead of recording a message, there would be an official White House statement that could be photographed and encoded into the fabric of the record along with the rest of the images.

'Carter was told pretty much what the other statements were like and who made them and that Waldheim's statement was there,' Frank says. 'And he decided he would make a statement ... but not speak it. And the reason was he thought that his strong Southern American accent would be an embarrassment to the Earth creatures. So he didn't want his voice on the record ... It's a nice voice, but he's got a strong accent. He spoke like a farmer and he thought this was going to be an embarrassment to the earthlings.' Sagan helped President Carter prepare his statement.

'Carl had a big input – and that has not been revealed almost anywhere, so far as I know,' says Frank. 'Carter's statement was a mix of thoughts from Carter, made into nice rhetoric by Carl.'

The message, officially dated 16 June 1977, is slightly wordier than the UN chief's. It gives a little introduction to the civilisation that built the Voyagers, summarising America as a community of several hundred million human beings, within a global population of, at the time, four billion. And the meat of the message, like Kurt's, is really rather good, the highlight being this section: 'This is a present from a small distant world, a token of our sounds, our science, our images, our music, our thoughts and our feelings. We are attempting to survive our time so we may live in yours. We hope someday, having solved the problems we face, to join a community of galactic civilisations.'

The White House, like the UN before it, would eventually put out a press release, telling reporters what the President had done and what he had said, and the overwhelming response was positive.

Cool. That wasn't so bad, was it? No indeed. Alas, there was more...

The Golden Record team were dotting Is and crossing Ts. But now the NASA machine was becoming increasingly hands-on and, according to Carl writing in *Murmurs*, since securing the President's message, there was 'concern' from NASA officials that the supporting tiers of government should also be represented in the record. As every good American schoolchild knows, Congress and the President share power, so shouldn't they both be on the record?

There was obviously not the time or space on the record for every member of the US government circa 1977 to say 'hello', but NASA wanted somehow to memorialise the existence of senators and representatives, particularly those members of committees who held sway over NASA activities and, as part of that, had a hand in the Voyager project itself. The result of this wish? *Four* photographs of typed lists of names. Names of members of the senate, names of members of the House of Representatives, and more specifically members of committees and subcommittees involved. So if a bespectacled alien researcher one billion years hence is wondering whether Max Baucus served on the House of Representatives' subcommittee on HUD-independent agencies, all it need do is pop the Golden Record onto some kind of playing device, set it to revolve at roughly 16rpm, get to the section encoded with pictures, download and decode those images using some kind of hardware that has the ability to transform sound to image, and there you have it. Four pages of names, among which – sandwiched between Bob Traxler and Louis Stokes – it will find the name of Max Baucus, and that will confirm that he did indeed serve on the House of Representatives' subcommittee on HUD-independent agencies.

This insistence on including a typed list of politicians came late in the day. The photographs, so carefully and lovingly selected by Frank, Jon and the rest of the image team, had by now been converted into audio format. The tape was ready to go to the record team in New York ahead of final cutting and pressing. Good old Valentin Boriakoff – our lovely sandwich-biter from image #82, who had already helped resolve many a knotty problem and was, by all accounts, a very positive 'how can we make this work' kind of guy – was about to save the day again. Val met Carl at the NASA HQ in Washington. Carl handed over the President's message, and the list of members of Congress. Val went to an ordinary suburban commercial developers and had the sheets turned into 35mm slides. The White House was planning to release news of President Carter's message to the press and they did not want any leaks. So while Val didn't have a briefcase handcuffed to his wrist, he did have to be present for every stage of the developing process so no one could sneak off any copies. Once he had the slides, he flew with them to Denver.

Again, Frank writes of how the team at Colorado Video would bend over backwards to help. At short notice, they accommodated this last-minute frantic dash to convert the presidential message, and the names of both House and Senate members of space-related committees. What wasn't recorded in *Murmurs* was that poor Wyndham was by now on vacation, celebrating his 28th birthday: 'I received a phone call at home, claiming to be from the White House, apologising for interrupting my time off, but requesting my urgent return to Colorado Video to reassemble the entire set-up and drive back to Honeywell to borrow the recorder again.'

At the same time another member of the National Astronomy and Ionosphere Centre at Cornell, an engineer named Dan Mittler, spent several 'days of inconvenience' shuttling between Boulder, Ithaca and NYC as they attempted to get this thing sorted. Once the bonus images were converted, the team asked Mittler to fly with the tape and the Honeywell recorder to CBS New York. As recounted in *Murmurs*, they couldn't risk putting the Honeywell in the hold, only for it to get lost in transit – there was no time. The airline had no option for allowing larger pieces of baggage in with the passengers, so they simply booked two seats, one for Dan Mittler and one for 'Mr Equipment'. And as Mr Equipment could truthfully be described as being under 10 years old, he was able to travel at half fare.

Judd says: 'It didn't mean too much to us at the time, few of us even knew who Carl Sagan was. Yes, ultimately it gave us a sense of pride and accomplishment. However, the pushiness of some of the people involved and little thanks given for a pro bono task, didn't leave us with an overly warm feeling. If we had known that it all had to be done in something like 12 weeks with almost no budget, we might have been more understanding and appreciative. Certainly dragging Wyndham back for a second go at it was not appreciated.'

The whole Pioneer vulva debacle still attracts a lot of debate, and yet in lots of ways this part of the Voyager story seems to me much more ridiculous. I don't mean to pour scorn on the creators of the record, the NASA officials who made it all happen, the structure of US democracy, nor this last-minute scramble to get it done. The road is paved with logical intentions, and each decision, when placed within its context, is understandable. But I just mean: *look!* Examine the practical upshot of this decision to include those four pages of names. Humankind had this unique chance to send anything – *anything* – of our world to the cosmos, and yet we chose to fill up part of the records' finite space (albeit a tiny part) with a printed list of utterly meaningless names – meaningless to an alien audience, at least. It should serve as yet another cautionary reminder of just how bad the Golden Record *could* have turned out had Carl and the rest of the team not managed to operate without input from outside agency, without letting politics muddy the water too much.

That said, as with every single atom of the Golden Records, the fact that converted images of a few pieces of paper did make it aboard our two celestial chariots is as illustrative to the Voyager record as the missing vulva is to the Pioneer plaque. We can scrub up, put on our best suits and flash our white teeth to the celestial camera, but we can't completely disguise our foibles, insecurities and idiocies.

★★★

At CBS it was time to carry out pre-flight checks ahead of final mixing and mastering. They were polishing up the sound essay with kisses and sped-up EEG readings, they were putting UN delegates to whale song, they were mixing the Cornell greetings, they were wrestling with multiple formats – from domestic reels and studio masters to commercial vinyl discs. And at last, by now a little late, the picture sequence arrived. However, there was a snag, and it looked like quite a serious snag: it was too long. Really rather too long.

Tim says: 'There were never any raised voices or anything like that. It was a beautiful project and there was really no reason to get angry about it ... When the imaging team delivered their data tape and it was twice the length that they were given on the record, and their response when I pointed this out was "well, just take some of the music off" ... I remained pleasant.'

Tim was determined that he was not about to make any cuts in the music to make room for the photographs. No way. So he knocked off for the night: 'I gave myself overnight to think about what to do ... That evening it occurred to me that, since we were going to be mastering in stereo, we could put half the photographic data in one channel and the other half in the other channel, which would bring the project back into compliance without reducing content. Crosstalk was better than -30db, so the solution was technically sound. I told Carl about this solution that night or the next morning.'

The final Voyager photographs reel-to-reel tape, which would soon be transferred to the lacquer masters and still survives in Sony's archives, is dated 10 June 1977.

Meanwhile, the final pieces of the playlist were still falling into place. One of the late runners for the Russian berth was 'The Young Peddler' performed by Nicolai Gedda. The suggestion came from Carl's composer friend Murry. It certainly ticked the Russian box, but they had nagging doubts. For a start the soloist was Swedish, born in Stockholm to a half-Russian father. The song was about someone seducing a young woman too, and it was felt this might not be the best way to represent the USSR. And as the Cold War was still at full tilt in 1977, they had to tiptoe around all things USSR. Carl cabled a Russian colleague, telling him what they were up to and which song was currently in the running as the Russian entry. Time was short and his cable included a deadline for a response.

Alan Lomax had already flagged up an excellent piece called 'Chakrulo', a Georgian chorus, which was in the running and they had a copy. Everyone liked the sound of it, but there was a snag with this too: they had no idea what the Georgian chorus was singing about. What if it was rude? What if it had some political message? It could have been about anything, and they were wary of something that might later come to light and embarrass NASA's good name once it was too late to change anything.

They need a Georgian. Lomax tracked down a man named Sandro Baratheli who lived in Queens. By this time it was the morning of the day the lacquers masters were due to be cut. Writing in *Murmurs*, Tim describes how Sandro came to the studio, listened to the song, lit a cigarette and began telling his desperately clock-watching audience all about the history of folk music in Georgia. At any other time, they probably would have been entranced, but there were engineers waiting. The masters had to be cut and then immediately transported to LA to be made into the final metal discs. Let's imagine rapid smoking, finger-tapping, furtive glances, clock-checking, beads of sweat appearing on temples, fingers pulling at shirt collars. Eventually, Sandro gave them the low-down: there was no problem with the chorus, in fact the lyrical content was inspirational – it was all about a peasant protesting against a tyrannical landowner. It was in.

Weeks later, far too late to make any difference to the record, Carl would hear from his Russian correspondent. It turned out that the request for a song to include had been taken seriously and had been passed up the chain to the top bods at the USSR's Academy of Sciences. The result? 'Moscow Nights', penned in 1955, originally as 'Leningrad Nights', a very popular and well-loved song. Writing about it in *Murmurs*, Carl and Tim were relieved they hadn't waited. While the song is a lovely thing, it's also a bit too standard, too obvious, rather than interesting or challenging. It's postcard Russia, tourist-board Russia. It represents Russia in the same way that 'Now You Has Jazz' from *High Society* represents jazz, or 'Poppa's Blues' from *Starlight Express* represents blues.

All the music was in place. Now they had to sort out which tracks went where.

***

They sequenced Earth's cosmic compilation in pretty much one night. It was late in the process, and they sat down and thrashed out an order. Tim thinks this was done at CBS Columbia by Ann, Carl, Linda and himself, right at the end of the sessions, just before the record was to be cleared by CBS.

'My one suggestion,' he says, 'that I had learned from other producers, was see if you can create pairs and triplets. Don't worry about the overall sequence yet. Just what two things go well together, what three things go well together. And then let's see if we can't build a sequence out of those units. That'll move this along faster.'

And it went pretty well. Most of the sequencing was done then and there in one long meeting. Ann Druyan kept notes.

NASA's compilation is a 27-track global tour, with a Western classical bias. It starts brightly with the first of three pieces from J.S. Bach – the Brandenburg Concerto – followed by Indonesian gamelan, percussive 'Cengunmé' and joyous 'Alima Song', before the sounds of clapsticks and didgeridoo recorded on Milingimbi Island, Australia, 1962. Two late-fifties floor fillers – the distinctive Mexican duel-trumpet of 'El Cascabel' and 'Johnny B. Goode' – are followed by a minute-and-a-half of alternating notes in 'Mariuamangi', then comes the Japanese shakuhachi piece, Belgian violinist Arthur Grumiaux performing Bach's second entry, before soprano Edda Moser sings Mozart.

The mood downshifts with two-and-half-minutes of haunting Georgian chorus, a minute of Peruvian drums, and three minutes of sleepy low-tempo perfection from Louis Armstrong. The Azerbaijani 'Muğam' sits ahead of a foundation-shaking slab of Stravinsky, a third and final piece from Bach, and the de-de-de-daaaaah of Beethoven's Fifth.

Valya Balkanska's powerful 'Izlel E Delyu Haydutin', is followed by the incongruous transition from a 1942 Navajo night chant to the Elizabethan 'Fairie Round'. Then comes the breathy 'Cry of the Megapode Bird', 'Wedding Song', 'Flowing Streams', before the record is brought to a close with the raga, Blind Willie Johnson and Beethoven's farewell flourish: the Cavatina.

Tim says: 'I always felt that "Dark Was the Night", which was the first piece I proposed to the record, would be good at the end or near the end. And the transition between that and the Cavatina is quite amazing. There are some great transitions, from my standpoint anyway.'

Ann says they gave a lot of thought to the last four or five pieces: 'If you look at David Pescovitz's reissue, his facsimile of the Voyager record, you will see included in the booklet a picture of my setlist in my own handwriting ... I wanted the last pieces to be night music. And, you know, just to have that sense of Voyagers' very long wandering...'

It's true that the record leaves our planet not with a celebratory fanfare, but with a tremendous sense of loss, of romantic longing, from the vocal cords of Kesarbai Kerkar, the wordless moans of Blind Willie Johnson, and the heart of Ludwig van Beethoven. The final piece, the Cavatina, was performed by the Budapest String Quartet at the Library of Congress in Washington DC in the spring of 1960. This was towards the end of the Quartet's career, which from around 1940 right up until 1967 made a series of landmark recordings with Columbia Records, finally calling time on public performances in the late 1960s when its three oldest members, first violinist Josef Roisman, violist Boris Kroyt, and cellist Mischa Schneider were all suffering poor health. By then they had set the standard for chamber ensembles, touring Europe, the Middle East and the US, and had recorded the complete Beethoven cycle for the CBS label.

Now, quite rightly, you're not interested in *my* opinion of all this. Who cares what I think about music? I have terrible taste. I'm known for being a little too partial to novelty music and throwaway pop, and was once, at the very last minute, stripped of the honour of DJ'ing at a university event when the organising committee overheard my boasted intention to begin my set with the 1988 novelty record 'John Kettley is a Weatherman' by A Tribe of Toffs. With all that in mind, I have a couple of things to say.

Firstly, considering that an image of a nuclear explosion was left off the picture sequence for fear of appearing like a threat, I find it surprising that the sacrificial dance segment of Igor Stravinsky's *Rite of Spring* made it aboard – and, by all accounts, made it quite easily. I say this with respect to the composer, to all the composer's fans, and to Murry Sidlin who suggested it. I understand the arguments for its inclusion, and I can see that it does have an interesting, complex and very striking structure. But it sounds to me like a species looking for trouble. It's about the most intimidating piece of music I've ever heard.

Now the Cavatina, on the other hand, is captivating. It's utterly, spellbindingly, stupendously beautiful. It's a piece of music that keeps promising to give you a note your whole body craves, only to give you another you didn't know you needed. It is the 'I Am the Resurrection' of the Voyager Golden Record, a triumphant slab of awesome that makes you yearn for something you can't taste or see, like a half-remembered, sunny summer-afternoon orange juice poured by your mum. It makes you remember things you've left behind, people you've hurt, and people you've loved, and people who you still love and yet are forever ageing before you and are so changed from the simple, easy people you first knew. It gives form to your past – a figure who approaches quietly, taps you on the shoulder and gives you a look you can't interpret. It makes you want to find anyone you've ever said a cross word to, and not say sorry, or say anything, but just urge them to sit down with you and listen to *this*.

With the emotional fallout that was on the horizon for the team, it is perhaps an appropriate piece of music to end a record that would last a billion years – forever imprinted in its final bars, the simple, daily difficulty of being human, of being without a muse, of being alone, drifting towards a last fork in the road. Beethoven never witnessed a performance of the piece in its final form, as it was first premiered after his death in 1827. And as Tim writes in *Murmurs*, it comes from the mind of a man who longed for marriage and family, but was refused because, according to Magdalene Willmann at least, he was 'ugly and half crazy'. The man, dubbed 'Immortal Beloved' by his fanboys, who wrote a famous letter declaring love for another human being – a letter he never sent. Ferris writes about the mood of the Cavatina, and how Beethoven's companion of his later years, Charles Holtz, said the composer could be moved to tears just by thinking about it. And how Beethoven wrote the word '*sehnsucht*' in the margin of his work, meaning longing or pining.

Some years before the events of Voyager, Ann heard the Cavatina for the first time. She was so moved, she wondered then how she could ever repay the gift of this piece of music. Voyager, it seemed to her, had given her exactly that chance. Talking about it four decades on, she says: 'In the Bible it's

40 years that the Israelites were in the desert. And now it's been 40 years that Voyager has been moving, at 40,000 miles an hour for every hour of those years. And yet that is the tiniest fraction. It's a nanosecond of a nano-fraction of the time ahead. And so I was imagining, how can we make a statement at the end of this record? The idea of endless night. The final statement, the coda, the final moment of the record, I wanted to be that sadness and solitary feeling. At the same time there's this little thing of hope in the Cavatina, that is to me so triumphant in what makes it so powerful. It's an acknowledgement of the struggle of life and how lonely and how difficult it can be. And yet the joy at the heart of it.'

I asked Jon how he reflects on the mixtape. For him, although he thinks they did a great job, he can't stop but think of it from the alien perspective: 'This is music that is important to us,' he says. 'We don't know if you're going to like it, but *we* like it. Maybe you will, maybe you won't. If it had been entirely up to me, I would have leaned much more towards ... things that you would do that you wouldn't necessarily need to do for a human audience unless you were teaching music appreciation or something. But that for an extraterrestrial audience it would be more like a class in Earth music. Maybe having the same tune played in different styles on different instruments, for example ...

'Whereas, as it is, it's more a sampler of Earth's music. Which is fine and I hope that some of the pieces at least will be understandable. I think that all the music, even if it is very culturally referenced to something particularly human, will still have built into it the elements of tempo and harmony and counterpoint that will be of interest. And I think that the nature of musical instruments will also be possible for them to understand – especially the acoustic instruments. I think those sounds – drums and flutes and vibrating strings and the human voice – especially given the information we supplied about our biology and what animals are like, I think they'll understand that some of these sounds are made by the bodies of animals.

'Where I part company with Carl is that he felt that …
some songs, some of the pieces expressed a kind of cosmic
loneliness. I mean, nobody knows anything about
extraterrestrials. But for what it's worth, I'm sceptical that
cosmic loneliness is as easy to convey as prime numbers.'

The last thing I'd say about the records' music is this: if the
Voyager records had turned out differently – say, if Frank's
massive noggin hadn't been available, and if that in turn had
meant that for some reason all the images and greetings had
taken up most of the runtime – and by June 1977 there was
room for just one piece of music, I would have voted for the
Cavatina. If one piece must represent Earth, and if it can't be
'Misalliance' by Flanders and Swann or 'Soulshake' by Peggy
Scott and Jo Jo Benson, let it be the Cavatina. It's really, *really*
good.

CHAPTER TEN
The Final Cut

'So let me see if I get this straight. You made this piece of shit and it's going to go whirling around the Earth, and for some reason I've gotta clear this motherfucker?'

Anonymous

Vladimir Meller was born in Czechoslovakia. He graduated from a high school in Košice, went on to study electrical engineering in Prague, learning violin on the side. Then, in 1968, Russia invaded.

'I was studying in Prague. I was in my third year. I went back to school and Russians were occupying pretty much every building, every police station, every newspaper building. I mean, it was just all over town.'

Vlado went back to his dorm to find most of his friends gone. He asked where everybody was: 'I hear, one guy's in England, one's in Rome. And I was like: "How in the hell did they get there?" Because you can't travel in communist countries. You just can't. You can't get a passport.'

The gossip was that the border was open between Slovakia and Austria. The Russians were watching the West German border – they didn't want people crossing over to West Germany – but Slovakia to Austria was clear. Vlado returned home and told his parents that he wanted out, that he wanted to escape. At first they thought him crazy, but he described the ghost school he'd returned to, and eventually they relented.

'So my father took me to Bratislava, the capital of Slovakia, and that's only 30 kilometres across from Vienna. At night you can see the lights in Vienna. That's where the border was open. We got to the train station and there were thousands – I mean, *thousands* – of people who were running. Families, young kids … And they were all trying to get on the train to Vienna. So my father put me on the train. I mean, you only have to cross the Danube River – that's the border. And you have Slovak guards on one side and the Austrian guards on the other side. And when the train stops at the border because you have to show your passports, nobody – I mean, *nobody* – on that train had a passport. Nobody. We were running from invasion – where do you get a passport? Who's got time to get a passport?'

The Slovak guards, perhaps worried about being overwhelmed or sparking an international incident, waved them across.

'As soon as we got to cross the Danube River the Austrians stopped the train. They were asking for the passports too. Nobody has passports. So they got us off the train and put us on a bus to Vienna and that's where my journey started.'

This was the autumn of 1968. Vlado was 21. He made it to Austria, and then by April 1969 he had reached the United States, legally as a refugee. He was given asylum, along with many thousands of Slovaks, Hungarians, Poles and Romanians all fleeing the situation in Europe. Then just months after arriving, in December 1969, he tucked into a bit of the American Dream, starting a new job in New York.

'It was unbelievable,' he says. 'It was a beautiful job, beautiful salary, beautiful benefits. I mean, I was in seventh heaven. You hear about all this stuff … "Jobs in America", people say. And there I was: 22 years old and I have a job.'

I know what you're thinking: who is Vladimir Meller and why are we hearing about him? Well, Vlado is the man who cut the Golden Record.

★★★

The sequencing and mastering was now complete, the audio tracks all on master tape, but before Vlado could start work, everything had to be cleared. The process of copyright and clearances was also handled on site by Columbia Records. Tim had thought it sounded risky – leaving all the copyright clearances until the last moment like this – but he had been assured long before that it wouldn't be a problem. So he walked into a meeting with the head of clearances for CBS Columbia and all his assistants. Tim was the only outside representative of the Voyager record project there.

'I started in on my first sentence to describe what we were doing and this guy interrupted me. "So let me see if I get this straight. You made this piece of shit and it's going to go whirling around the Earth, and for some reason I've gotta clear this motherfucker? Why? Why would I do that?" And I said: "Because your CEO told you to do it and he told you to have it done by close of business Friday." After that we got along fine.'

They cleared all the tracks in about two working days. Then they took the tapes finished in the editing department to Vlado in the cutting room.

Today Vlado is an audio-mastering engineer who runs his own business in Charleston, South Carolina. His career credits include metal, rock, hip-hop, jazz, opera, pop and classical. He's worked with the Beastie Boys, Johnny Cash, Michael Jackson, Limp Bizkit, Paul McCartney, Metallica, George Michael, Oasis, Pink Floyd, Public Enemy, Rage Against the Machine, Red Hot Chili Peppers, Shakira, Weezer, Kanye West and Jack White. CBS Records first hired Vlado in December 1969 and he worked for the company until going solo in 2007. In 1977, he was in his thirties and was the top cutter at CBS.

Vlado told me: 'They came up to me, introduced themselves and they told me what they want to do, and what the record would be about and that it will go into space … Obviously, as a young dude working, I'm thinking: "These guys are space cadets … who's going to play this record?" Obviously, I'm in a different position now. I realise it was an unbelievable project. I didn't realise when I was cutting it that it would be such a big deal 20, 30, 40 years later on.'

The role of the cutter was to take the two-track master, put it on a tape machine, and start cutting into the vinyl using a cutting lathe. Next thing you know, you have a disc in your hand that you can play back and it plays music the same way as it sounded on the tape.

'I remember it very well,' says Vlado, 'because it was almost impossible to cut that record. It was an awfully long record. I mean it's difficult to cut even 28-, 29-minute records, never mind 35-, 40- to 45-minute records, because the grooves are so packed on the vinyl, so thin that even with your best cutter and best engineer there's a possibility it could skip. Conventional turntables just can't track it because when it starts spinning the arm is being pulled to the side ... a super-thin groove, you have a chance of skipping.

'So we had to do lots of test cuts. Then play it back on my regular turntable at CBS Records and make sure that it played all the way through. Even though there was lots of spoken word, where the grooves are not modulating very much and you know they're at low level, but then there was lots of music also. And once the music comes in and there's lots of classical music ... then you start getting modulation and the grooves are spreading apart from each other. And it can eat up the space on your record very quickly.'

Despite working at 16rpm and using a state-of-the-art computer-driven cutting lathe, they were stretching the audio brim, pushing the technology to its limit. If the grooves became too thin it just wouldn't work. To inscribe the photographic data into the lacquers – remember they were using both sides of the stereo grooves – they plugged 'Mr Equipment', the Honeywell recorder, directly into the cutting board to minimise distortion.

'This combination of noises, and bird noises, and whales, and this, and animal noises, and spoken word ... You could save some space on the record for music – you could then open up the grooves for the music. So it was very creative how we changed the levels from each depending on what the source was. If it was a spoken word we could lower the level slightly and save space, if it was music we could crank up the levels slightly so the playback was decent ... But we cut it, we did it, we played it. We were safe. It played all the way through. It was an unbelievable achievement,' he says.

There was one more job. Engineers and cutters signed off vinyl records by etching matrix numbers issued by the record company into the run-out groove – that bit on the surface between the end of the grooves and the circular paper label in the centre. Some were in the habit of leaving messages in the run-outs too, such as legendary Blackburn-born engineer George 'Porky' Peckham who from the 1960s frequently signed his records 'Porky' or 'A Porky Prime Cut'. John Lennon, as Tim knew from working with Jimmy Iovine, also had engineers etch messages into those blank spaces, so Ferris thought: why not use the space for something? It was another opportunity, another stretch of canvas. He asked Vlado to add a dedication, one originally written a full month earlier on the back of an envelope during one of the listening sessions. It is, says Vlado, 'a very delicate process. One bad move and the master would be ruined.'

The hand-etched inscription reads: 'To the Makers of Music – All Worlds, All Times.' The best thing about this inscription though, is not the meaning of the message, but the words themselves, how they appear on the surface of the record. Hold them up to the light, and they look like normal, casual, everyday handwriting, like a 'back-in-five-minutes' message scrawled on a piece of paper. It's the only organic visual element to the records, a fingerprint, a tiny artisan touch, delivered to the universe by Tim Ferris and Vlado Meller from a cutting room in New York, and which, a billion years from now may be the only trace of anything hand-crafted in the cosmos.

Reflecting on his work some 40 years later, Vlado sees it all as a stroke of good fortune. 'It was an unbelievable project. I never worked on a project like that in my life. I worked on lots of rock 'n' roll, classical, jazz, this and that, but Voyager was something very different. Carl mentioned me in his book, which was very nice of him. But then you have all these articles on the internet and they don't even *mention* the cutting at CBS Records. They talk about the records being cut by Colorado JVC – it's fuckin' nonsense! They had nothing to do with it!'

★★★

By this point, what with all the last-minute greetings, presidential messages, overrunning images, relaxed Georgians, and a truly challenging cutting process, the deadline had expired. Luckily it turned out that John Casani, like all good editors, had kept a few spare days hidden away in his schedule.

Vlado finished cutting at 11p.m. Minutes later, Tim left the CBS building with freshly cut lacquers under his arm. He took a cab directly to the airport, caught the red-eye to Los Angeles and hand-delivered the discs to the James G. Lee Record Processing Center in Gardena, California, the next morning. Here the two sides of the disc became two, single-sided copper mothers (the blank discs provided by the Pyral S.A. of Creteil, France). In the place where a normal record would have its paper label, there was to be a photo-engraving of the Earth viewed from space. On the record it looks like a fairly indistinct swirl of clouds. And finally over this photograph are printed the following words:

THE SOUNDS OF EARTH
Side 1
NASA
UNITED STATES OF AMERICA
PLANET EARTH

These words, coupled with the 'music makers' inscription, are the only words on the outside of the record itself. If it is discovered but for some reason never fully decoded, those are the words that will represent our epoch: all, America, Earth, makers, music, NASA, of, planet, side, sounds, states, the, times, to, worlds, united, 1, 2.

Each record is in fact two one-sided copper mothers bonded together. First they were gold plated, then joined together to form one 0.05-inch-thick Golden Record, weighing in at around 1.25 pounds.

Then they did it all over again, to make the second Golden Record.

★★★

One of the most mystifying parts of the Golden Record, especially to us humanities majors, is the cover. As Jon wrote: 'To the average earthling this whole diagram may seem obscure.'

Frank had already outdone himself with the Pioneer pulsar map, but now he faced a bigger challenge. This was design at its most pure: something brief that would be understandable to beings with no known common ground, aside from that offered by mathematics and physical laws. The cover had to communicate not only where Voyager had come from, but also when it had come from, what was inside the circular metal box, how it worked, what it meant, what you were supposed to do with it … all within a blank space about the size and shape of a record.*

In the upper left-hand corner are top and side views of a phonograph record. Expressed in Frank's binary transition-of-the-hydrogen-atom time unit, is the correct time of one rotation of a record played at the correct speed. The designs show how the record needs to be played from the outside in, and there's also a side view of the record and stylus with another binary number giving the time to play one side of the record. Below that is the pulsar map. Then we swing around to the upper-right portion. This explains how to

---

* An often-forgotten person in the story of the cover is Barbara Boettcher. Back in 1977, she was a drafter at Cornell, who transformed some of the plans, diagrams and sketches into perfect, architectural-standard blueprints.

reconstruct pictures from the recorded signals. The top drawing shows the regular wave form of the video signal that occurs at the start of a picture, which, if correctly decoded, traces a series of vertical lines. Then comes a diagram explaining the audio duration of each picture. Below that we come to two rectangular shapes that between them explain all about picture raster, how the 512 vertical picture lines should be redrawn with staggered 'interlace', and showing the correct 4 × 3 aspect ratio of the resulting image. The lower rectangle contains a perfect circle, the final piece in the picture puzzle that will correspond with the first image in the sequence.

Jon writes: 'The fact that each block of information consists of 512 signals of identical length, and that each of the 512 signals is only slightly different from the one before it should be a clue that the signals are lines that are to be stacked. Recipients may try to stack and reproduce these in various ways, but as soon as they reconstruct the circle it should be evident that they have done it right, since the chances of randomly producing a perfect circle are small, and they will recognise from the cover that a circle is what they are supposed to produce.'

Remember there's no separate 'how-to' for the colour images on the record's cover. There wasn't room and it would have been far too complicated. Plus it was thought that as the colour images were represented in the audio-code by three sets of sound in sequence – one each for the red, green and blue components of the image – the aliens would figure out there was something different about *these* sound images.

Finally, there are two circles in the lower right-hand corner. This is the all-important hydrogen key – the hyperfine transition of neutral hydrogen. These represent the hydrogen atom in its two lowest states, with a connecting line and digit 1 to indicate that the time interval associated with the transition from one state to the other is to be used as the fundamental timescale, both for the time given on the cover and in the decoded pictures.

Jon writes: 'This is what happens when hydrogen's single electron changes its orbit around the atomic nucleus. Under stimulation by an energy source like starlight, the electron jumps up a level, and then falls back down … As it does so it emits a photon of a particular frequency and wavelength. This will seem a very esoteric event to most readers of this book, but it is in fact the commonest event that happens in this universe, since space is filled with hydrogen atoms absorbing and re-radiating the light from surrounding stars. Radio astronomers listen for this radiation as a main probe in mapping the universe. If other civilisations know anything about physics, they will know about this energy transition in the hydrogen atom. If, and this is a very big if, they recognise our diagram of this transition, we will have given them a unit of length and a unit of time.'[*]

The aluminium covers were etched at Litronic Industries, Irvine, California. They also added the uranium clock – electroplating a small area of each cover with an ultra-pure sample of the isotope uranium-238, which, as it has a half-life of 4.468 billion years and a known rate of decay, would give the recipients a rough idea of the record's age. The record with the cover, support and mounting, weighed about 2.4 pounds. The needle and cartridge, the electronic components required to transfer information from the stylus to an electrical wave, were mounted to the underside of the support.[†] Closer to launch, the records would be fixed to the Voyagers at the John F. Kennedy Spaceflight Center at Cape Canaveral. They were attached with side one facing inwards, where there is a slightly reduced rate of damage from cosmic rays, and considerably reduced risk of pitting from micro-meteors.

---

[*] Remember that the unit of length is the wavelength of the energy hydrogen emits, about 21cm. The unit of time is how long it takes for this transition to occur – a tiny fraction of a second.

[†] Photographs of the final production process survive at the JPL Archives: voyager.jpl.nasa.gov/galleries/making-of-the-golden-record/.

'This was done,' says Tim, 'so that if the record were to be recovered more than a billion years into the future and the outer side was badly eroded, the surviving side would still contain all the photos, the sounds of Earth, the greetings, and the first tracks of the music.'

Everything was ready to go. The record team were beginning to breathe more easily, returning to mundane, everyday work. They were looking at old to-do lists and in-trays, unchecked for six weeks, and looking forward to launch with the knowledge that the Golden Record – *their* Golden Record – would be aboard.

However, there was a last-minute, heart-quickening, bureaucratic snag that almost sunk the whole project just as it was cleared for take-off. Although NASA had seen and heard the contents of the record and given its approval (barring the naked pregnant couple), the object itself still hadn't been officially signed off. Just as it seemed Carl could put his feet up, he received a phone call telling him that this precious metal mixtape, into which he had poured six months of his life, had been rejected.

A by-the-numbers NASA quality-control officer was checking the record against specifications. In other words, he had a form on which was written the expected physical characteristics of the Golden Record, and he was checking that against the real thing. If it had been badly overweight, for example, it would have thrown all NASA's thrust vector calculations wide of the mark. It was found that, while the record's size, weight, composition and magnetic properties were all in order, its blueprints made no provision for Tim and Vlado's 'to the makers of music' inscription. There was nothing on the form about any kind of handwriting. So the record was rejected as a 'nonstandard part', and the space agency actually began preparing to replace it with a blank disc. Carl made some frantic calls, persuading the NASA administrator to sign a waiver, allowing the records to fly.

★★★

Ann and Carl took the four-hour Circle Line cruise around Manhattan to talk things over. This was about a week after first declaring their feelings on the telephone. They were obviously happy, delirious even, but also consumed by guilt. It was pretty clear that a scandal could jeopardise the record. A NASA uneasy with nudity would definitely be uneasy with the record's celebrity figurehead abandoning his wife and family. People had begun to realise something was going on. Carl's personal assistant Shirley noticed an American Express bill that showed Carl paying for Ann's transportation. Shirley drew this to his attention and Carl admitted he had fallen in love. Ann's friend and confidante Lynda Obst, meanwhile, had encouraged her to go full *Anna Karenina* and to follow her feelings.

Wendy remembers it too: 'It was palpable and profound. Back then, as a young person figuring out relationships, and now with the passing of time, I understand that I was witness to the beginnings of something rare in my experience; that is, two people with an unusual and deep connection that cannot and frankly, should not, be denied, stopped or ignored. Life is short. And tragically, it was very short for Carl. I observed Ann and Carl's relationship unfolding in front of me – their rapport and synergy could not be missed.'

By the end of the Circle Line cruise, they had mapped out a journey. In the distance was being together and, while there were hurdles, obstacles and collateral in their path, they knew that was the place they wanted to go. They would keep everything quiet for now. They made an agreement that they would break the news to Linda and Tim at the same time – at 1p.m., two days after the Voyager launch.

CHAPTER ELEVEN

A Last Supper

'Jeez. We got away with it.'

Tim Ferris

Another secret was about to get out. That irritating UN press release had already spilled some Voyager beans but, in the main, the record project had continued without attracting too much unwanted attention. Carl had wanted to keep it that way for several reasons, not least interference. As the aftermath of the record's launch would show, the moment people knew about it, everyone had an opinion. If it had been public knowledge during the compilation, imagine the influx of lobbying, leaning and hectoring. It might have been overwhelming.

Nevertheless, in late July – the Golden Record complete and now just a few weeks before its official press launch – Carl received a call from *Wall Street Journal* scribe Jonathan Spivak. Carl describes the call in *Murmurs*. Spivak already seemed to know a lot of the details. And Carl kept his responses within the parameters of confirming what the journalist already had, without giving him anything else. There was obviously some confusion along the way as the final piece, printed in the *Wall Street Journal* on 26 July, included a reference to Duke Ellington being on the playlist.

This article forced NASA's hand, who put out a more comprehensive press release ahead of schedule a few days later. A collection of images, letters, notebooks, files and ephemera from Shirley Arden's working archive, sold in 2017 by Boston-based auctioneers, included a copy of this 'For Immediate Release' document. The headline reads: 'Voyager will Carry "Earth Sounds" Record.' It begins: 'On the chance that someone is out there, NASA has approved the placement of a phonograph record on each of two planetary spacecraft being readied for launch next month to the outer reaches of the solar system. The recording, called "Sounds of Earth", was placed Friday (July 29) aboard the first of two Voyager spacecraft scheduled to be launched to Jupiter, Saturn and beyond. The 12-inch copper disc contains greetings from Earth people in 60 languages, samples of music from different cultures and eras, and natural sounds of surf, wind and thunder, and birds, whales and other animals. The record also contains electronic information that an advanced technological civilization could convert into diagrams, pictures and printed words, including a message from President Carter.[*]

It's interesting that NASA chose to lead this press release, dated 1 August 1977,[†] with that 'Earth Sounds' headline. Perhaps they felt this made it sound more sober, more scientific, more grounded in what NASA was all about – rather than the more frivolous content of 'Johnny B. Goode'. Indeed, it doesn't mention Chuck Berry at all, although it does give a pretty full account of the rest of the record, detailing its physical characteristics, along with the greetings, the languages, the different genres of music, and the presidential message.

More newspaper coverage followed – most of it positive. And pretty soon, as had happened with the Pioneers, hundreds of letters began to arrive at NASA and on Carl's desk at

---

[*] You can read the rest of the original press release via the JPL website: www.jpl.nasa.gov/news/news.php?feature=6047.

[†] Just around the time NASA was finalising the Golden Records for alien consumption, our planet received an unexplained radio signal. The 'Wow! signal' was received on 15 August 1977 by Ohio State University's Big Ear radio telescope. Astronomer Jerry R. Ehman spotted an anomaly in the printed readings, circling the area and adding the comment 'Wow!' – hence its widely used name. The signal has not been detected since but remains an acknowledged but disputed candidate for alien radio transmission. It seemed to come from the constellation Sagittarius, although many believe it was a natural phenomenon.

Cornell. Most of these were enthusiastic, although as before, there were some kooks horrified that the record team's activities might give away our position to attack-minded aliens.

A posed NASA photograph taken on 4 August 1977 shows John Casani standing in front of Voyager 2, holding a modestly sized American flag in his hands, the 12-inch gold-plated copper disc and the reverse of its aluminium protective jacket laid out on a white sheet. This was taken at the John F. Kennedy Space Center, Cape Canaveral, just before the record, cartridge and needle were fixed to the side, the flag folded and sewn into the thermal blankets of the spacecraft.

Days later the entire record team gathered in Florida to witness take-off.

'When we got to the Cape the first NASA staffer to walk up to me scolded me for not putting "Danny Boy" on the record,' recalls Tim. 'He said that a "fine Irish lad like myself", knowing that Tip O'Neill was the Speaker of the House, should have put "Danny Boy" on the record. Well, what can I say? I love "Danny Boy". You know, I play "Danny Boy" myself. But what can I say? It's not on the record.'

For the launch they gravitated towards the NASA bleachers, the distant Titan Centaur rocket poised on its pad. Carl, Linda, Ann, Jimmy, Tim, Frank, Jon, Amahl, Wendy ... Tim's mother Jean Baird Ferris was there too.

Frank says: 'The whole earth shakes even though we are a mile or so away from the spacecraft taking off but you just have this sense of enormous invincible energy being exhibited.'

Jon writes: 'The countdown. An explosion of smoke and steam, vast billows pouring out from the base of the distant rocket. The vehicle began to rise. I felt myself rising with it, gathering speed, dwindling to a point in the sky, then gone. It was as clear and pure a moment as I have ever experienced.'

Sagan also wrote about the emotion of the day, how they smiled and wept to see it leaving the Earth. At one point, amid cheers and hugs, Carl shook Jon's hand, congratulating him and thanking him for all his hard work.

Tim too recalls talking with Carl: 'It was largely relief,' he says. 'I think a lot of what Carl and I were saying to each other at that moment of watching the rocket go up was: "Jeez. We got away with it. We actually did it." Because the whole record was in peril at one point, and there was plenty of anxiety about it. We were really happy that it was actually happening. Launches are always emotional anyway. I'd seen launches before. I'd gone to the Cape to see them when I was a high-school kid. I'd seen them flying by from Key Biscayne, up on the roof of our home to watch rockets flying by at night. So I felt at home with rocket launches. And I was kind of amazed that I'd had something to do – however small a role – with a mission that flew. Having at age 12 watched these things go off, I was really gratified to have even a tiny role in a mission – that was emotional too.'

The record team had dinner in a restaurant popular with NASA staffers. At one point a tipsy Italian-American NASA press officer tottered over to the table. He said: 'You put *three* German composers on the record and not *one* Italian one?'

Jon writes: 'He gave us a gesture of such forceful clarity that I wish we had put a photo of it on the record as an example of how humans communicate non-verbally.'

The man had a point, of course. There were three Germans in Bach, Beethoven and Mozart, but no Corelli, Monteverdi, Verdi or Vivaldi. Jon says: 'Anybody who loves a genre of music that was not represented has every right to feel irked that what was special to them wasn't included. And I totally sympathise. Which is just an argument for doing more of these things, I guess.'

People disagree about what NASA actually thought about the record. It's certainly true that Sagan detected an air of detached bemusement at some of the sounds and pictures when it was all cleared at NASA back in June. In the 40th-anniversary documentary, *The Farthest*, Frank talks with a mischievous twinkle about how the record always got more attention than the Voyager science, and perhaps that's true more recently. But certainly, it wasn't always like that, and Jon Lomberg, speaking to me from his home in Hawaii in 2017, said that what Frank described in the film wasn't *his* experience at all.

It's certainly possible to think that NASA didn't really know what they had. The record is nothing if not strange. And while there was no pesky vulva staring them in the face this time, there was now so much more on there that could potentially attract criticism or cause embarrassment. In any event, looking back, Tim felt the press launch for the record side of Voyager was at best rather apologetically managed and at worst completely mishandled. He felt it was NASA trying to hide that the record even existed. He noted that in nearly all the press materials – images, diagrams – the Voyager craft was shown from an angle that didn't show the record. Initially there wasn't even going to be a press launch for the record at all, but as some reporters kept on asking about it, at the last minute NASA scheduled a room at Frank Wolfe's Beachside Motel.

The official Golden Record press conference took place in the afternoon, four days after the death of Elvis Presley, and the day after Groucho Marx died. This was a historic moment – the only time in the six-month gestation of the Voyager record when the entire record committee were together in the same room. They had rubbed up against each other as sub-groups – at Carl's house, or the Smithsonian, or CBS, or Cornell – but in the main they had operated as individual teams. Now they faced the press corps together, and it would have been fascinating to know the tone of questions. Did reporters ask about specific songs? Did they quiz them about Chuck Berry? Were they interested in how the President's message came about? Did they touch on the printed sheets of politicians' names? Did any of them scratch their heads about the whole 'putting a photograph on a record' thing? Sadly we can never know as no audio or video of this press conference survives.

Talking to me in 2017, Tim called the press conference 'a joke'. The motel, which has since been torn down, was double-booked. The record's official launch was held in a large room separated from the oompah sounds of a full-blown

Polish wedding[*] by nothing but an accordion-style folding
barrier. Thanks to this sound pollution, no tape of the Q and
A has survived.

In *The Farthest* you see fleetingly some photos from the
press conference. Jon is seated next to Frank, who is dressed
in a yellow shirt, a pen sticking out of his breast pocket. Carl
is in an open-necked white shirt tucked into fawn trousers.
Ann is wearing a dark blue shirt and dark jacket, seated
between Carl and Tim. Tim is wearing a dark blue shirt,
cool-ass glasses and is drinking a coke. Frankly the team look
a little tired and gloomy, but you can't assume too much from
a couple of stills.

I asked Tim if he remembered feeling nervous.

'No no no. I always felt that everybody was going to like
the record. I thought that the only issue was with NASA –
and I understood NASA's concerns. I thought they were
making a mistake by trying to basically hide the fact that the
record was even on the spacecraft. But I was perfectly happy
to be presenting, you know, to be answering questions to the
press. I thought the public was going to like the project. I've
always felt that the record would be popular if it were ever
given a chance. I've always welcomed public interest in the
project … I was just disappointed that NASA had done such
a poor job. We said a lot at that conference and there are no
recordings of any of it because the audio people couldn't work
in that environment – with a wedding reception taking place
on the other side of the accordion divider. And whether that
was suppressed hostility on NASA's part or just an honest
mistake, I don't know. But of course if they'd scheduled a
proper press conference in the first place then there wouldn't
have been issues. But that's it, you know? With Carl pushing
the envelope into areas outside NASA's orbit, naturally you
get some push-back when you do that.'

★★★

---

[*] He tells the story in *The Farthest* too, about 20 minutes in.

After the press conference, Tim and Ann drove to Miami, and Carl, Linda and Nick drove to the airport to fly to Cape Cod.

Ann had confided in Lynda; Carl in Lester Grinspoon. Together they decided that Carl should tell Linda at the Grinspoon's place in Cape Cod so the Grinspoons would be on hand to provide moral support as friends of the couple. On 22 August, two days after the Voyager 2 launch and 82 days after the 'for keeps' phone call, Carl told Linda he wanted a divorce and that he wanted to marry Ann.

Ann says: 'We had made this vow that at one o'clock we would turn to our significant other and tell them what had happened and that we were in love and that we were leaving. We both of us did it at one o'clock, unfaltering, never looked back, never faltered. And that was it.'

According to Poundstone's book, Linda was shocked, angry, distraught. She'd wanted a second child with Carl. Arguments and discussions went on into the night. Eventually they drew up a list of their personal items and possessions on a yellow legal pad and began to divide things up then and there.

Tim, again according to Poundstone, took it fairly calmly. He had already experienced heartbreak and knew the ropes. And amid all of this upheaval there was Nick. The only child aboard Voyager. A sweet little six-year-old who had said 'hello' to the universe for all of us kids just a few weeks before. He too is reported to have taken the news in his stride.

Just around this time, Sagan's own parents had moved into a retirement condo in Florida. Carl met with his father, Samuel, telling him that he was leaving his wife for another woman. Samuel said, without looking round, that he hoped the woman was Ann.

Linda and Carl's divorce would run for some time, with days in court and coverage in the press. Friends and allies helped out on both sides. Frank visited their house in Ithaca, witnessing Linda and Carl splitting up their possessions. In the immediate aftermath, Linda didn't want Ann and Carl being seen around Ithaca together. So while they didn't immediately move in together,* they did travel and in December 1977 they visited London.

In December Carl was invited to the White House to brief President Carter and the first family – essentially to give them a kind of celebrity astronomy lesson over dinner. Carl did that, and then the next day he and Ann flew to London to prepare for the Christmas Lectures.

The Christmas Lectures have been held at the Royal Institute in London since they were started in 1825 by Michael Faraday.† Still broadcast every year, they are a series of lectures on a single subject to a general audience, but aimed at younger people. Carl Sagan was a perfect choice for the Christmas Lectures – a scientist with a sprinkling of celebrity cache, like David Attenborough (who did his in 1973) or Richard Dawkins (who did a particularly good job in 1991). Carl was sandwiched between Nobel Prize-winning chemist George Porter and knot-theory mathematician Christopher Zeeman. He had formally been invited to deliver the lecture on 14 May 1976 by Royal Institute Director Sir George Porter.

You can go online and watch the videos now. In the first lecture Carl imagines what an alien visitor might deduce from looking at our planet. Then he turns to the outer solar system, to life, to Mars and the prospect of planetary systems beyond our sun. In the first lecture, he mentions the just-completed Golden Record, and plays an excerpt from 'Depicting the Cranes in Their Nest' over the address system to a spellbound audience. If you watch very carefully, you sometimes see fleeting glimpses of Ann sitting in the front row, amongst the pleasingly geeky-looking crowd of late-1970s youngsters. Ann, like the rest of them, is watching Carl.

Tim went back to work. He returned to his writing, taught English at Brooklyn College, City University of New York,

---

* They first lived together from January 1978 – a home in Slaterville Springs on the outskirts of Ithaca.
† Aside from a wartime break between 1939 and 1942.

and launched his first book, *The Red Limit: The Search for the Edge of the Universe*, which won the American Institute of Physics Prize. In a later profile, Tim described his life during this late-1970s period as being that of a 'happy bachelor'. He would meet the woman he would settle down and have a family with in 1979. Although his relationship with Carl was tarnished, a mutual respect remained. In his acknowledgements for *Coming of Age in the Milky Way*, a monumental work he wrote between 1976 and 1988, he acknowledges debts of gratitude to both Ann and Carl (among many others). He dedicated one of his own documentaries to Carl Sagan and, when speaking to me in 2017, he still talked about Carl with great warmth.

★★★

Opinion is divided about how much attention the Golden Records got at the time. We've lived with the achievements of the Voyager missions for decades. We've watched this great age of discovery unfold before us. But at the time of launch, all those discoveries were still in the future. So it must have seemed like just another NASA probe launch to many writers and readers. Indeed, if you put yourself in the boots of the contemporary press corps, it *was* just another probe. It must have sounded quite a lot like the Pioneer probes. The fact that these ones had LPs strapped to their sides was the unique feature, at this point in history at least.

The Sunday edition of the *Idaho Times* covered the launch on page seven, after a full-page obituary of Groucho Marx on page six. The story mentions a 'gold-plated recording' in the intro, reports on some early technical gremlins encountered at launch with one of the gyroscopes and the camera-carrying boom, before ending with a brief rundown of the record's contents – touching on the human kiss, the Louis Armstrong tune, Beethoven's Fifth and the sound of an automobile. It also seems to indicate that NASA had provided an actual turntable for the record as well – which they didn't, of course.

The *St Petersburg Times* covered the Voyager 2 launch with a tiny paragraph, just above the corrections, and alongside more coverage of the deaths of Groucho and Elvis. This merely noted that it was taking along with it 'information about Earth' in case anyone is out there – although it makes no mention of the record itself.

The *Cornell Chronicle*, dated Thursday 25 August 1977, alongside announcements of the forthcoming 'Freshman Offbeat Olympics' and new laws reducing penalties for marijuana use, proudly proclaims that Voyager 'Takes Disc Aloft', with 'Sights, Sounds of Ithaca Included'. The accompanying piece brilliantly draws out the local angle – that the record includes the voices of '27 Ithacans', plus photos of the produce counter at the Grand Union Supermarket in the Cayuga Mall and the Cayuga Heights interchange on Route 13.

Over time, as the Voyagers sped further and further away, the records occupy less room in the limited column inches. The *Virgin Islands Daily News*, on 6 March 1979, reports on the first photographs beamed back from Jupiter, in what a NASA spokesperson calls 'one of the most memorable days in planetary exploration'. This report ends with a little Golden Record primer, reminding readers about the discs, telling them that once the Voyagers' primary missions are done and dusted, they will be carrying their precious 'phonographs of Earth' into deep space on the off-chance of an alien encounter.

Then by the 1980s, as Voyager 2 made its triumphant fly-bys of Uranus and Neptune, there are times when the record hardly gets a look-in. If you only have five paragraphs to summarise newly discovered geological features on a Jovian moon, or drifting clouds of debris around Neptune, or brand-new never-before-seen planetary rings, or faint radio signals given off by Uranus, suddenly the Cavatina is *way* down the pecking order. So during the 1980s science and exploration was the story, and the records – while never completely forgotten – were mere footnotes.

Jon says: '[Frank] said* that during the mission the press was much more interested in the record than they were in the science, and that annoyed the scientists. That's just not true. And Frank was not at any of the Voyager encounters that I know of. He's not a planetary astronomer. So I don't know where he got that. It might have been a bit later, his own experience. But as somebody who covered the mission for the CBC – and I was at every encounter from the first Jupiter one to the final Neptune one; I was a known member of the press corps, friends with all the other press – in all that time I don't think anybody *ever* asked me about the record. There was no sense of "hey, it's Jon, he worked on the Golden Record." Nobody cared about the Golden Record. First of all, at the beginning of the mission, nobody knew that the mission was going to be as successful and as remarkable as it was. Nobody knew the solar system would be as incredible as it was. So there wasn't the sense of historic monumentality of the mission until much later. And the record was a side thing. I don't recall the record being discussed in any of the press conferences except maybe the very final one – at Neptune. And I don't remember reporters asking any questions about it. So I think it took a while to gain traction.'

Pop-culture appearances helped it gain a measure of mythic status. There was Steve Martin on a contemporaneous *Saturday Night Live* episode, reporting on the first message received from aliens (it reads: 'Send more Chuck Berry!'). In the first *Star Trek* film, in 1979, Kirk and Spock discover a menacing alien force calling itself 'V'ger'.† In the early

---

* In *The Farthest*.
† Wyndham Hannaway, the man who converted the Voyager photographs to sound, took a leave of absence from Colorado Video in 1978, working at Robert Abel and Associates on motion control photography of a six-foot model of the *Enterprise* for the film. It wasn't until he saw the premier a year later that he discovered that the *Enterprise* crew was up against the mysterious 'V'ger'. 'Astounding,' he says, 'but definitely not my script idea.'

1980s Jon received a treatment for a never-to-be-made film called *Stella*, in which an alien responds to the record by assuming human form and moving to New York. Then there was the flawed *Starman* film* in 1984, in which an alien sent to Earth in response to the record's invitation crashes near the home of a bereaved widow (Karen Allen), takes the form of her former husband (Jeff Bridges), and first communicates with a deadpan rendition of Kurt Waldheim's opening remarks. In 2016 there was 'A Glorious Dawn', a piece of autotuned spoken word by Sagan that Jack White had pressed into a gold-plated vinyl 7" and played on a 'space-proof' turntable suspended by a high-altitude balloon. And more recently the record plays a prominent role in the young-adult novel *See You in the Cosmos* by Jack Cheng.

'I think these helped to lift it – give it the lustre that it has now,' says Jon. 'There probably hasn't been a week that has gone by in the last several decades where I haven't gotten some kind of message … So I hear about it. I know it's out there and people love it. Stage plays are being written about it, an opera has been written about it, a symphony has been written about it. So it's kind of made its mark. But from my point of view, the fame and glory associated with it … it's like one of those jokes about the stupid person's lottery. It's a million dollars – it's a dollar a year for a million years. The fame and glory is, for me anyway, associated with the record, like that. It hasn't made me payments and it hasn't made me rich. But I think as long as we remember the space programme we're going to remember Voyager. And as long as we remember Voyager we're going to remember the Golden Record.

'I mean, I have no celebrity status, I have no name recognition or brand. I was at JPL a few weeks ago – for a screening of *The Farthest* – and the woman is giving us a tour of JPL. There in the auditorium is this full-scale model of the Voyager, and there's the Golden Record on

---

* In the film Charles Martin Smith plays Mark Shermin, a SETI scientist who put the record together. The film is also responsible for spreading the misconception that the records are broadcasting the song 'Satisfaction'.

it. I went up to her afterwards and I said: "You know, I helped make that cover. I was the designer of the record." She said: "Really? What's your name?" I said: "Jon Lomberg." She said: "Lambert?" She'd obviously never heard of me.'

CHAPTER TWELVE

Hello, We Must Be Going

'Carl saw our society, our planet, the life on it with the kind of … the perspective of someone from another world. One of my favourite things said about Carl was: "With terrestrials like Carl, who needs extras?" There was a deeper truth there which was that he did have that … cosmic perspective.'

Ann Druyan

We had a record player in the sitting room of my home. It was up high, on a wooden shelf, so toddlers like me couldn't mess with it. Below the Garrard turntable was another shelf with an amplifier (a Leak Stereo 30 Plus), and below that a small cupboard with a modest collection of vinyl.

The speakers were set into the wall. They had been put there by my grandfather, a DIY audiophile and opera nut who used to live in the house where I grew up. Inside the cupboard, just next to the records, was a line of four dome-shaped, Bakelite switches. These had been installed by my grandfather too and could be used to switch on speakers secretly mounted in various positions around the house. Flick one, and a speaker hidden behind a framed picture of a horse in my parents' bedroom burst into life. Flick another, and a speaker in the bookshelves of a downstairs room began to hum. This may sound fairly run-of-the-mill to the wireless Bluetooth generation, but back then this was pretty cool.

I was too young to touch the records. I wasn't trusted to handle them correctly, and anyway I couldn't have reached the turntable even if I'd wanted to. So if I wanted to listen to Penelope Keith reading fairy tales or Shakin' Stevens singing 'This Ole House', I had to ask for parental help.

I remember listening to the powerful crackle as whichever parent I'd coerced to my will dropped the needle to the groove. I remember the building excitement as I waited for the smooth clear vinyl to give way to the spikes of rich analogue sound. I'd listen to stories like 'The Steadfast Tin Soldier' or 'The Tinderbox', often alone, rocking back and forth on my haunches.

However, I was terrified of another sound. I dreaded it. I never knew when it was coming, but it always did in the end. And when it came, I would run from the room as fast as I could, absolutely terrified.

Turntables often have some kind of built-in mechanism that kicks in when the playing arm has reached the end of the record, automatically lifting the needle, returning it to the rest position. Not so at our place circa 1980. Ours would just go on, spinning round and round, the tone arm and needle endlessly skipping back, 33⅓ times a minute, against the end of the run-out groove. The powerful amp and the bass-rich wall-mounted speakers meant that, to my young ears, that sound of needles skipping over the run-out was the stuff of nightmares, conjuring up images of heavy-footed monsters. *Whoomph … che-t-ke … whoomph … che-t-ke … whoomph … che-t-ke … whoomph … che-t-ke … whoomph … che-t-ke …*

Even today when I listen to records I still feel the same shiver when I'm too slow to lift the tone arm before the record reaches the end. It's a desolate sound. The sound of something approaching. The sound of an ending.

★★★

From the day Voyager 1 joined its sister ship in space on 5 September 1977, the records that were attached to the sides began to fade from popular memory as the important business of the missions – the planetary fly-bys – took centre stage.

I mentioned at the start of this book that for me Voyager 2 was like a hero. A spiky space insect that looked nothing like spaceships in my mind's eye or in films, and yet, every now and then, would broadcast back all this amazing news and pictures from places we'd never been. Just as part of my brain watches a record spinning and still can't really figure out how all those tiny ridges make music, I still look at photos of the Voyagers and can't really believe they did it – and are still doing it – these strange, ungainly vehicles that weigh about the same as a small car and (if you removed all the weird spikes and booms – the magnetometer boom is the really long one at about 13m) would fit inside a 4m² box.

These robotic explorers would give us many marvels to wonder at in our libraries and living rooms. They not only informed Sagan's own *Cosmos*, but their discoveries formed the bedrock of the BBC *Planets* series in the 1990s, the planetary *Brief Encounters* series of the 2000s, Professor Brian Cox's *Wonders of the Solar System* in 2010 and the most recent *Cosmos* reboot manned by Neil deGrasse Tyson. So let's take in some of the highlights of exactly what it is they *did* do.

The closest approach to Jupiter occurred on 5 March 1979 for Voyager 1, and on 9 July 1979 for Voyager 2. The closest approach to Saturn occurred on 12 November 1980 for Voyager 1; on 25 August 1981 for Voyager 2. Primary mission complete, Voyager 1 headed off. Voyager 2 had survived its Saturn encounter, so NASA applied for and secured that extra funding needed to run the bonus tracks on the Voyager LP – and it headed off to Uranus and Neptune. But there was to be a gap of nearly half a decade before Voyager 2 came to Uranus (closest approach: 24 January 1986), then another gap before Neptune (25 August 1989*).

Between them Voyager 1 and 2 explored all the giant planets of our outer solar system: Jupiter, Saturn, Uranus and Neptune. They either flew by, studied, measured or photographed around 50 moons. They buzzed Titan, studied the gravitational shepherds that shaped Saturn's intricate icy rings. They discovered new planetary rings and studied magnetic fields. They found ice worlds and oceans, moons tortured by gas giants flexing their gravitational fists, geologically active moons, moons with thick gloopy atmospheres, ice moons, moons with geezers, all of them bathed in perpetual twilight.

---

* NASA's New Horizons probe crossed the orbit of Neptune 25 years to the day after Voyager 2's encounter. It reached Pluto in July 2015.

They introduced us to all kinds of interesting characters along the way. Look at Miranda, for example, one of five sooty nutters circling Uranus, and one of the oddest objects in our solar system, with a huge palisade known as Verona Rupes, cliffs more than twice the height of Everest. Between them, the Voyagers found more than 20 new moons. They made the first measurements of the magnetospheres of Uranus and Neptune, took the first detailed images of the rings of Uranus and Neptune. They found active volcanism on Jupiter's satellite Io right at the start of the planetary encounters, ending the final fly-by of Neptune with the discovery of yet more active geezers on Neptune's smaller, colder moon Triton. They also found an unexpectedly epic storm system swirling on Neptune, with winds exceeding 1,000mph – the fastest recorded in our solar system. And they did all this with computers on board that couldn't power your smartphone – an 18-bit Computer Command System (CCS), a 16-bit Flight Data System (FDS), and an 18-bit Attitude and Articulation Control System (AACS).

A few years after the dust had settled on the Neptune fly-by, there was to be a final flourish. JPL planned to shut down the cameras on Voyager 1, which hadn't been close to any significant bodies since the winter of 1980. Carl proposed that, before they switch them off, they turn them back towards the Sun to take one final photograph of the solar system. There was opposition from JPL. For one, this was easier said than done. Controlling something remotely from so many AUs (Astronomical Units) isn't simple. Plus, there was little or no scientific value to the idea and it would drain more of the vessel's finite reserves of power. Not only that, but the picture team knew that the photos would be very likely to turn out rubbish – home planets at that distance would be little more than points of light. Planetary scientist Carolyn Porco describes in *The Farthest* how Sagan kept pushing, eventually going up the chain of command at NASA, who eventually ordered JPL to comply.

The result has become known as the Pale Blue Dot photograph. It was taken on 14 February 1990, by Voyager 1, at an approximate distance of 3.7 billion miles or roughly 40.5 AU. Voyager 1 also took photos of the other planets, creating a kind of solar-system family portrait. And in the photograph, Earth is indeed a tiny dot, almost hidden in a band of sunlight. Carolyn talks about how when she first saw it, she assumed it was a glitch.

Look at the photograph and you'll see: there we are, suspended in a sunbeam. Sagan describes us as a 'mote of dust'. He unveiled the photograph at a June 1990 Voyager press conference. He's introduced by Ed Stone and takes to the stage, looking older than in the halcyon days of the Golden Record, more considered and methodical in his movements and speech. He stands behind the NASA logo. What follows is among the most famous speeches of his career. With simple eloquence and tender humanity, he describes how that pinprick, that tiny speck, is *us*, is our home. He remarks that every single person you've ever known, or read about, lived on that dot. He shows how insignificant we all are, how tiny our home is in the vastness of space, and how this should teach us all humility. With a photo and a lectern Carl Sagan stands in a pale brown suit, and saves the world, revealing our 'fervent hatreds' and posturings for what they are: ridiculous, when placed in an interstellar context.

The Pale Blue Dot press conference informs a memorable chapter in the 2017 documentary *The Farthest*. It's a pin-drop, hair-raising moment from the final third of the film. You might think the emotion is being manufactured in the edit or by the soundtrack, but I watched the full press conference on YouTube. On comes Carl. He stands in front of a rubbish-looking TV, and this rather strange series of uninspiring composite photos appears. At one point they go one photo too far and he has to ask them to back up. Then he makes his speech (you can see the original typed draft in the Library of Congress). And guess what? The effect was *exactly* the same. For several seconds he holds the assembled room of jobbing hacks and reporters in the palm of his hand and, in a few well-chosen words, summarises the magnitude of a pixel, putting humanity and history in its place in the vast ocean of everything.

This is another part of Sagan's story where the cynical might ask: 'Did he have another book to sell?' But, for once, he didn't. Not then. Yes, he would title a later work *Pale Blue Dot: A Vision of the Human Future in Space*, but that was seven years hence. Indeed, the work closest to the Pale Blue Dot press conference was *A Path Where No Man Thought*, a 1990 co-authored book with Richard Turco, a gloomy exploration of what nuclear war could mean for humankind.

I'm sure this press conference didn't do his profile or book sales any harm,* but by now Sagan was using his position to champion certain causes – some through what biographers have called the 'Annie effect'. Certainly Ann Druyan encouraged him to use his fame for good. And he was a man concerned with the direction Western culture was going. He spoke and campaigned on women's rights, on the legalisation of marijuana, on climate change and the environment, on poverty, and the nuclear arms race. With this short speech, he cuts to the heart of the matter. By simply showing what a delicate speck we are, he shows how our planet needs to be cherished, appreciated and cared for, not abused or taken for granted.

This scientifically insignificant part of the mission has become a celebrated chapter of the Voyager story. It shows just what a tremendous gift Sagan had for spreading awe-inspiring ideas. The anniversary of the Pale Blue Dot speech was marked by NASA and the Planetary Society, and he would repeat tweaked and improved versions of the speech at talks and events for the rest of his life.

In February 1998, Voyager 1 passed Pioneer, becoming the most distant human-made object in space. In December 2004 Voyager 1 crossed the 'termination shock' into the heliosheath – 94 AU from the Sun. Voyager 2 crossed it three years later.

---

* Just to give you a flavour of Carl's stature: in 1983 his agent Scott Meredith struck him a new four-book deal with Random House worth $4 million.

In August 2012, Voyager 1 entered interstellar space – the region between stars – and today the interstellar phase of the mission is still going strong. Both spacecraft are *still* beaming back information about their surroundings through the Deep Space Network. We are *still* receiving readings from these amazing machines, almost half a century after their launch, with instruments aboard enabling technicians and astronomers on Earth to study magnetic fields, investigate low-energy charged particles, cosmic rays and plasmas waves. Both Voyagers are expected to keep at least one of their functioning instruments going into the mid-2020s. If they still have enough of their dwindling power – and assuming they are still within range of the Deep Space Network that picks up their signals – JPL engineers think we may still be receiving data from them in 2036. Right now you can go on the JPL website and see how far each vehicle is from Earth. Sometimes the distances seem to be going down – which is because the Earth moves around the Sun more quickly than the Voyagers are moving, so we catch up with them a little during certain times of year.

So they've gone a long way right? No, not really. If you took a diagram showing the solar system, with the solar plane tipped up at the back, slightly towards you, so all the planetary orbits appeared elliptical, and with Neptune's orbit now appearing to be about the size of the palm of your hand, the distance Voyager 1 travelled in a year would be about the width of your little finger. Even now, their journey has barely begun.

If you want to try and figure out exactly where they are right now in the night sky: Voyager 1 is heading north of the ecliptic plane towards the constellation Ophiuchus. In about 40,000 years it will come within 1.7 light years of AC+79 3888, a star in Ursa Minor. Voyager 2 is heading south towards Sagittarius and Pavo. In about 40,000 years it will come within 1.7 light years of a star called Ross 248, in the constellation of Andromeda. So long guys.

★★★

The Voyager records followed hot on the heels of Groucho Marx. And just as Groucho sang 'Hello, I Must Be Going' at Miss Rittenhouse's party,* it seemed like the records were doing the same – being introduced to the world, only to promptly quit our sphere forever: 'Hello, we must be going.'

From now on, any opinion about the record was just that – opinion. Nothing could be done to alter their contents. The records are humanity's tattoo. And like that tattoo you chose aged 18, which you've perhaps come to hide beneath the arm of your sleeve now you're 58, there might be things we'd change if we could, but we can't. Frank gave the example of Picture #72, the Olympic runners.

'I picked that one out,' he says. 'I thought it was a great picture. It showed we were athletic. It showed we had competitions. It told them we had different races on Earth. It showed we had spectator sports. And so I thought: "Well, this is wonderful. It shows a whole bunch of important things about us." But since then I've looked at, and if you look at it carefully, most people don't see this till I tell them ... Every one of them has one leg up. Really, you can't tell that it's a fluke of the photograph. You really could conclude that all four of them has one leg shorter than the other. And the other thing about it is they all are in the air – not one of them has a foot touching the ground. And so I have this vision of all across the galaxy people getting this picture and saying: "Does this mean they have an anti-gravity system on this planet? And they have this second species which is very athletic and has one leg shorter than the other?"'

We can't change the Golden Record, so it's best not to waste time with regret. Celebrate it. Cheer its triumphs, smile at the idiosyncrasies, and shrug at its missteps. Wear our 'Summer of '77' tattoo with pride and thank the tattooists who gave it to us.

Over the years, we have learned more about what's *actually* on the record than was known at launch. Check official histories of the Voyager project, or at least uncorrected

---

* See *Animal Crackers* (1930).

histories, and next to image #112 – a bold colour photograph of human figure on a spacewalk, face masked by a golden glass visor – is the name James McDivitt. McDivitt did indeed fly with the Gemini and Apollo programmes, and he did conduct spacewalks. But it isn't McDivitt, it's Ed White. White* was one of three astronauts who died in a fire during a test of the Apollo 1 Command Module in late January 1967. Midway through the test, a fire broke out in the pure oxygen-filled cabin, killing all three men. A year and a half before the accident, Ed took the first ever spacewalk by an American, on 3 June 1965. It was described as a 'spine-chilling 20 minutes' in the *NY Daily News* a couple of days later, a 17,500mph speed thrill, at 150 miles above the surface, on the third orbit of Gemini 4's four-day flight. The Gemini spacecraft was *piloted* by James McDivitt, and that day White surpassed the time spent in space by Soviet cosmonaut Alexei Leonov a few months before. Mid-spacewalk, a joyous Ed sang 'This is fun!' in the radio chatter, before Houston said something along the lines of: 'Ed, come on in here before it gets dark...' And this image, from a euphoric and joyous moment in Ed's life, is preserved forever in golden metal.

We've heard how the team decided early on that this was going to be a 'humanity on a good day' record – informative, encyclopedic, tourist-guide Earth. One of the reasons Wagner missed out on a seat was because of his association with Hitler and Nazi philosophy. And it wasn't just Wagner: Jon's 'Alleluia' suggestion was sung by Elisabeth Schwarzkopf, who at one time worked in Goebbel's Reich Chamber of Culture; and Herbert von Karajan, conductor of the nominal choice for the Beethoven symphony, was a former member of the Nazi party, so his celebrated version of Beethoven's Fifth was dropped in favour of the Otto Klemperer-conducted performance. You can imagine, therefore, how awful it must have felt – how embarrassing it was – when the record team found out in the mid-1980s that the very first voice you hear on the record, the very first sounds to come forth when

---

* Ed White was buried with full military honours at West Point.

needle hits metal groove on side one, issue from the mouth of
a former card-carrying Nazi.

The story broke around the time Voyager 2 was on its final
approach to Uranus. Waldheim, after finishing his term as the
UN Secretary General in 1981, was running to become the
ninth president of Austria (his term lasted from 1986 to 1992).
And it was while running for the 1986 election that his
wartime work as an intelligence officer in the Wehrmacht
came to light, sparking an outcry and investigation –
particularly into the part he played in Operation Kozara in
north-western Bosnia in 1942.*

I talked about this issue with both Frank and Jon. In short,
they both shrug and chuckle. *Typical.* You make all this effort,
put in all this work to present us humans as upstanding
citizens, and… 'What did we end up doing?' says Jon. 'Well,
the first voice you hear is a Nazi. And you know it's … It's
humorous in a cosmic sense. And when they did the kiss [for
the sound essay], Tim still thought Annie was his fiancée and
he would give her a kiss that would last for a billion years, but
the reality was that she's thinking of some other guy that she's
in love with. I mean, what is more human than that? You
know? I mean, it's almost operatic. That as far as we tried to
sanitise the message, some of the most primal human struggle
and basest and rawest emotions got in there anyway … You've
gotta laugh. It's a cosmic irony.'

In 1991 Tim had described how, broadly speaking, there
were three criteria at work when choosing music: geographical
diversity, economic diversity and 'good music'. So the
economic argument might be why they arrived at Blind
Willie Johnson, say, rather than Gershwin. The wish for

* The 'Waldheim affair' was broken by investigative journalist Alfred
Worm in Austrian weekly news magazine *Profil*. It centred around
Waldheim's claim that he received a medical discharge after being
wounded in the winter of 1942. Simon Wiesenthal, the Jewish
Austrian Holocaust survivor and Nazi hunter, stated that Waldheim
was stationed 5 miles from Thessaloniki when its entire Jewish
community was sent to Auschwitz in just a few weeks.

geographic diversity, it seemed, was checked with input from experts such as Lomax and Brown. But who decides what's 'good'? Perhaps my tone sometimes drifts towards snooty when describing tracks on the record I don't care for as much as others. And some people may listen to some of the tracks today and think, 'how did *this* get on?' Who the hell knows what is 'good'? The idea that some white intellectual New York lefties should be the arbiters of taste for a planet might stick in your craw. And all this might be an argument for siding with Jon Lomberg's view that the music should have been chosen purely by it having an interesting, examinable or mathematical structure. But maybe that too, if taken to its final conclusion, would have felt disappointing four decades down the track. Carl and the team wanted to share some of our emotions, our soul with the cosmos, and I think they achieved that goal.

Alan Lomax came to see the record as a wasted opportunity. This must have felt like a project built for him when it landed on his desk. Here was an opportunity to put his 'science' in space, to record a tiny litmus compilation of his vast musical database, his life's work, squeeze it into this golden bottle and lob it into the cosmic ocean. Whereas Robert Brown, reflecting afterwards, felt the whole idea of having any kind of satisfiable criteria for choosing the music to represent a planet was in itself a hopeless task.

The record has attracted criticism. People accuse it of naivety, of wearing a skin of cod inclusivity. You could argue the team seemed open to influence when it came to the non-Western parts of the mixtape, but a little more proprietary when it came to selecting their own music. You could also see it as little more than a publicity drive for Carl Sagan's own career. Indeed, if you do, it must go down as one of the most long-lasting, sophisticated and successful publicity drives in the history of publishing.[*]

Carl's marijuana use began around the tail end of the 1950s. There's nothing unusual about marijuana use, of course, but

---

[*] Three members of the record team had books out in 1977.

his open stance as a casual user and his on-the-record advocacy were relatively unusual, especially for an academic, albeit a populist communicator academic. It's another facet to his wide appeal, making him a figurehead and hero[*] to both mainstream and counterculture America. Indeed, he wrote a pseudonymous essay (as 'Mr X') for his friend Lester Grinspoon's 1971 book *Marihuana Reconsidered*. Sagan fervently believed his outlook and perception had been helped by marijuana use. He felt it helped him learn and understand new things he otherwise would not have learned or understood. He believed it had literally broadened his horizons, both in terms of his appreciation of art and through scientific epiphanies. It inspired many of his ideas, including some of the more out-there 'ideas riding'. But even at the time of the 'Mr X' essay, Sagan firmly disagreed that an epiphany he had when smoking weed would reveal itself to be worthless once given sober examination the following morning. The feeling of an idea when high was not the mere 'illusion' of great insight, he wrote. It was *actual* insight, but capturing and codifying that insight, pinning it down, recording it in some form that could then be taken forwards the following morning was the tricky part.

You could certainly make an argument for some marijuana-laced fingerprints being on the Golden Record. Indeed, its very existence is down to Carl's passion for sending music into space, which must, at least in part, have been powered by some marijuana-fuelled enthusiasm. The fact that they wrote about an alien race being able to one day decode Ann's thoughts from those firecracker EEGs is pretty down-the-rabbit-hole stuff. And it doesn't seem too much of a stretch to conclude that some of the music choices were helped over the line by that same enthusiasm. But of

---

[*] Carl's use was intermittent and moderate, rather than excessive. But it continued throughout his career. And later in life he became a more outspoken public agitator for the legalisation of medical marijuana.

course, you could say exactly that about a lot of 20th-century music and about a lot of 20th-century records.

Another point I'd like to make is a simple one: think about how *bad* this record could have been. I mean, there are any number of potential horrible outcomes that were avoided. Have you heard some of the popular chart hits from the 1970s? Don't get me wrong, I like a bit of 'Chirpy Chirpy Cheep Cheep', but I don't want the aliens knowing that. Worse still, what if it had been a really insipid *Best Classical Album in the World Ever!*-type compilation? Good music, sure, but how *boring*. The whole thing could have been absolutely strangled by committee, or could have become sponsored and commercialised, turned into little more than a record-label advertisement or back-catalogue sampler. As it is, we ended up with this multifaceted cultural artefact. A mixtape where a 2,500-year-old Chinese refrain plays alongside Japanese flute, where Peruvian panpipes vie with Navajo chants, an Indian raga and a Cavatina pluck at our heartstrings, where Mexican mariachis duke it out with a 1920s Chicago seven-piece, where the sounds of youthful euphoria rise from the Ituri Rainforest and the strings of Chuck Berry's guitar, all the while rubbing shoulders with Bach, Beethoven, Holborne, Mozart and Stravinsky.

'A lesser person – you see this all the time – would have instinctively gone to appointing a NASA committee in which the majority of the votes would have been NASA staffers,' says Tim. 'And I've served on some of those committees and they do get certain things done, but they would have been disastrous for something like this. So I think Carl's genius was to keep it within a small group where we could really do creative work, and not have to try to fit everything into some 11-dimensional social matrix that might or might not have offended anyone … NASA was scared to death of any kind of criticism coming from any member of particularly the House or Senate. I mean, just to the point you would describe them as clinically paranoid on that subject. So they would have endlessly been tiptoeing around, saying, "what would so-and-so think?" and "shouldn't we be doing this?" And we did hear *some* yappings like that in the distance, but I think we really got a good record because of the way we did it. It was a bottom-up matter of inspiration and intellect on the part of everyone involved. I don't mean that you couldn't have selected a dozen other people similarly who would have done just as well or better. But I do think it was brilliant not to get into some sort of bureaucratic nightmare. It is very hard to get creative work out of that … as any architect will tell you.'

I spoke to as many of the record team as I could when researching this book,[*] and it felt like there were areas of tension over ownership of the record – who was responsible for what. Talking to Ann, Frank, Jon and Tim, it felt at times like talking to proud parents, each with strong views about their child. And, as any parent knows, the person you can sometimes find most irritating in all the world when discussing your offspring is the other parent. Views differ on how things came about, on who was responsible for certain things at certain times. People remember things very differently.

'No, actually, she was with *me* when she lost her first tooth. Remember? It was while we were roasting marshmallows.'

'No, that was her second tooth.'

'No, it wasn't.'

'Was.'

'Wasn't.'

'Was.'

So there's some Rashomon effect at play here. At the same time, though, there's plenty of mutual respect, with an acknowledgement from all parties that everyone involved worked hard, did their best and contributed positively to a project that came out well. Let's just say that they don't often get together and go bowling. Which is a pity. I'd really like it if they occasionally got together and went bowling.

★★★

---

[*] I also reached out to Linda Salzman Sagan.

One of the great frustrations for the record committee was that earthlings never got to hear this music for aliens. Today you can very quickly find audio files in varying degrees of quality online, through the JPL Voyager mission archive or NASA's own Soundcloud channel. But that's only been possible relatively recently. For years, the possibility of an Earth-released Voyager record languished in development hell. For decades the only way you could hear the music on the record was by reading *Murmurs of Earth* and tracking down your own copy of the original songs – assuming they existed on vinyl, which many did not.

It was never given a proper commercial release, despite Carl's best efforts. There were a number of factors: rights issues, confusion,[*] intransigence, lack of effort. Jon remembers being accosted by one high-school music teacher who said: 'I demand to see a copy. It is my right as an earthling!' There was eventually a CD-Rom version, which came with a reissued edition of *Murmurs of Earth* in 1992 (made by Warner New Media, with guidance from Frank and Jon), but that wasn't really satisfactory.

David Pescovitz, a former student of Tim Ferris, was batting around ideas with his friend and record store owner Tim Daly. They were searching for something they could work on together, when they discovered their shared fascination with the Golden Record. So David approached Tim Ferris, asking if he would endorse a first proper reissue of the Golden Record in time for the Voyagers 40th birthday. Tim gave the project his blessing, but under two conditions: it had to be the entire record, and they had to source it from the original masters. He wasn't going to get behind anything half-arsed.

---

[*] In 1977 the team had assumed that Columbia/CBS would eventually release a commercial replica. However, there were endless permission problems since some copyrights were held by direct competitors. Plus, the company's marketing department felt record stores wouldn't know where to put it. It was too much of an oddball. Jon writes: 'We learned the hard way that it was easier to send the record across the galaxy than release it in the marketplace.'

David and Tim Daly formed Ozma Records (named in honour of Frank's famous experiment). They spent almost two years clearing all the rights, trying to track down the stories of some of the indigenous music, correcting errors and filling the numerous gaps. Having recruited Lawrence Azerrad, they mocked up their design for a proposed triple-LP set of golden-hued vinyl, featuring all the music and sounds, plus a companion book with all of the record's photos and other ephemera, and launched a Kickstarter campaign in 2016.

The crowdfunding effort went live on a Tuesday. By Thursday they had already surpassed the $198,000 funding target, after seven days had topped $650,000 with more than 5,000 backers, and by the end raised nearly $1.4 million.

During the course of the project they tracked down original musicians, sending letters and paying royalties, they donned white gloves to hold the original lacquers at JPL archives, they followed Lomax paper trails at the Library of Congress, they met with Frank to take new scans of old diagrams. They were painstaking and obsessive and treated the record and everyone involved with it, or recorded on it, with an infectious reverence. Speaking about it in late 2017, David said one of the things he and Tim Daly were most proud of was their corrected tracklist. Then in January 2018, at the 60th Annual Grammy Awards at Madison Square Garden, all this work was rewarded with the 'best boxed or limited-edition package' award.

The Ozma reissue records, just as Tim Ferris had hoped, were based on the original CBS studio masters. After the lacquers had been cut by Vlado back in 1977, CBS Records (now Sony) had deposited the original master tapes in a climate-controlled underground warehouse where they sat, untouched, for 40 years (the 1992 reissue was based on a domestic reel-to-reel copy). Then Tim and the Ozma team gathered at Sony's Battery Mastering Studios in New York City. An engineer named Vic Anesini – who had just baked the old reels to temporarily prevent the iron oxide from shedding off the backing – pressed play. David called the sound 'breathtaking'. Tim called it a relief: 'A great relief because we didn't know whether the tape was going to survive – with analogue tape any problems loom really large. If you lose half a second of analogue tape you've got a problem. None of those problems emerged. The engineer initially was starting and stopping the tape and logging each track. I pretty quickly asked him to stop doing that because I just wasn't even sure I trusted the backing not to snap. But it sounded terrific.'

There was another interesting little coda from the Ozma story. During the publicity push, a man named Ron Barry approached David. He said he had experience in video encoding, and asked David if he would send him the audio readout of the Voyager photographs so he could try decoding them using only the instructions on the case. David sent him a digital file of the audio, taken from the original master tapes. What happened next can be seen on YouTube, in a mysterious, multi-framed video where the rasping white noise of the picture sequence soundtracks the ghostly appearance of each image in turn, shown in real time as they appear simultaneously via the right and left channel.

'I was very interested,' says Tim, 'because … we had tested this technology early going into the record. Then there had been a test on the outside, of the data coming out. And I never felt that the images had been decoded properly. I felt they were *encoded* properly, but I didn't think the decode test was correct. To my eye we should have had about 30 decibels better signal than we were seeing. But this guy came back, and they looked great. It was fantastic. It was indeed the readout that was in error back then, not the record itself. The photos on the record look amazing.'[*]

★★★

---

[*] You can see the YouTube video here: youtube/ibByF9XPAPg. And read Ron Barry's description of how he did it here: boingboing. net/2017/09/05/how-to-decode-the-images-on-th.html.

So, what happened next? Tim met the future Carolyn Zecca Ferris in 1979. They were married in 1985. She had two children from a previous marriage, Francesa and Alex, and together they have a son named Patrick. They live in North Beach, San Francisco, and Rocky Hill Farm where they built an observatory in 1993. Tim has authored 12 books, from *The Red Limit* (1977) and *Coming of Age in the Milky* Way (1988), which won the American Institute of Physics Prize and was nominated for the Pulitzer, to *The Whole Shebang: A State-of-the-Universe(s) Report* (1997). He's made documentaries *The Creation of the Universe* in 1985 and *Life Beyond Earth* in 1999 (which ended with the dedication 'For Carl Sagan'), and more recently has written, produced and narrated *Seeing in the Dark* in 2007. The latter closes with the sound of Mark Knopfler playing 'Dark Was the Night' on a 1932 National steel guitar.

Tim would go on to teach at Brooklyn College, the University of Southern California, California Institute of Technology and UC Berkeley (where David Pescovitz was one of his students). He also nearly went to space as a finalist in the 1986 NASA Journalist in Space programme, which was suspended after the *Challenger* crash.

Speaking to Jon in 2017, he described how he still sometimes feels like he's back in the room at Cornell's Space Sciences Building, searching for the perfect image: 'I could not find a good picture of a campfire. People sitting around the campfire playing music, cooking marshmallows, talking, laughing … couldn't find one. Then of course a week after launch I found the perfect one. For years afterwards, right up until recently, I'd be looking through a book or something and I'd see a picture and think: "Boy, that would have been perfect. Where were you when I needed you?"'

Jon continued collaborating with Carl for the next 20 years, illustrating books, working on the *Cosmos* series, designing the original sailing-ship logo for the Planetary Society that Carl founded in 1980, and working on the film version of *Contact*. He's settled in Hawaii with his wife and two children. He's had an asteroid named after him and created 'Galaxy Garden' (the Milky Way mapped in plants and flowers, at Paleaku Peace Gardens Sanctuary in Kona, Hawaii). He created numerous artworks for NASA, he co-designed the 'MarsDials' artefacts – sundials placed on the Spirit and Opportunity Mars rovers inscribed with the words 'Two worlds, One sun' – and created time-capsule-like messages for future Mars explorers in the 'Visions of Mars' CD-ROM/DVDs placed aboard Phoenix, which landed on Mars in 2008.

When I spoke to him, a Kickstarter campaign he launched had just failed to reach its funding target. Jon had sought to recapture some of the Voyager magic, this time crowdsourcing images from people all over the planet and using them to form the basis of a new digital message to the cosmos that he hoped would be remotely uploaded to the New Horizons spacecraft (which had gone without any kind of message).

'I think we've lost faith in the future,' he says. 'I think we're afraid of the future. I think the future has become much more dystopian. I think that, for all the risk of thermonuclear war in 1977, people were more optimistic and positive. We were going to go to this wonderful *Star Trek* future of global harmony and interstellar exploration. And then the future became *Terminator* and zombie apocalypse and terrible plague and computers taking over and … I mean, every future is worse than the other. So I think we're in a very sick culture. And I think we're in a culture that has a very pessimistic view of itself and of the future. I view the response I've been getting to this project as being a symptom of that. Maybe that moment of history and that small group of people were just some unique never-to-be-repeated moment that allowed humanity to do something like that. But I don't think we're going to do it again. I see no signs of us doing it again. I mean, we could have done it again and didn't. They didn't put something on New Horizons and if they don't go for the current project I'm pushing, and there's no real indication that they will … Either you let the fifth spacecraft to leave the solar system go with a message or without it. That's your binary choice. And so far they're choosing to leave without it. And what does that say?'

Jon was such a positive force on the Voyager record, it seems a pity to say goodbye to him on a downer. So instead I'm going to quote a paragraph from the unpublished manuscript he was kind enough to share with me. Let's leave him in better spirits.

'One of my strongest recollections of the process was how well the team worked together. I think we were all a little intimidated by what we were trying to do, and that tempered the giddiness of the emotions. In a sense humility grounded us, something very necessary when you are reaching higher and farther than anyone has ever done before. People had strong opinions, but nobody was dogmatic about them. There was a sense of great responsibility, jointly borne. Carl was on the phone dealing with NASA; Frank was converting the periods of the planets into binary multiples of the wavelength of hydrogen; I was sifting through piles of *National Geographics* looking for a farm scene; Eck was shooting slides of photos I had already found; Tim and Ann were in Washington searching out the sound of freight trains; Linda was trying to track down someone who could say 'hello' in Swahili; George and Val were on the roof photographing the Sun's spectrum; Amahl, Wendy, Ros and Shirley were keeping everything organised. Whatever dissensions and dramas later split some of the team members from enjoying amicable relations with one another, while we made the record we worked together superbly. I hope the men and women who craft future interstellar messages are as lucky as I was in their choice of teammates and leaders, with whom I am proud to be cruising through space until the end of time.'

★★★

After suffering from myelodysplasia for two years, Carl Sagan died of pneumonia at the age of 62, at the Fred Hutchinson Cancer Research Center in Seattle, Washington. It was the morning of 20 December 1996.

In 2006 Ann Druyan wrote an article entitled 'Ten Times Around the Sun Without Carl'. She writes about what her life with Carl was like. She recalls the 'exquisite June day' on the Circle Line boat, just days after declaring their love for each other, planning out their future, looking back at him as they disembarked, a dazzling smile across his face as he throws a sweater in the air in exultation. She describes a man who would always stand up and shake the hands of fans who approached them when they were eating in restaurants. She writes about driving around the Ithaca countryside in Carl's orange Porsche, seven-year-old Nick in the back. She describes Carl playing with their own children, Sasha and Sam, and a Y-shaped twig handed to Carl by toddler Sam, which Carl then carried with him for the rest of his life. The eulogy, which you can read in full at anndruyan.typepad.com, ends with an acknowledgment of the gulf in distance and time, of 'ten long trips' around the sun since she last saw Carl's smile, but also with the sense of celebration at the life they shared.

Carl and Ann were married in June 1981 at the Hotel Bel-Air in Los Angeles. Shirley Arden was the matron of honour. Carl wore a pale suit with a dark tie, Ann dressed in traditional white with plenty of frills. As she walked down the aisle to marry Carl, a string quartet played the Cavatina. They lived in California for two years during the production of *Cosmos* in the late 1970s into the early 1980s, but most of the time their home was Ithaca.

They were never really apart after that. Sagan was profiled in *The People* in 1980. At that time he was basking in the afterglow of the records set by *Cosmos*,* the spin-off book had entered the *New York Times* bestseller lists at number one, and they were working on a treatment for the story that would eventually become *Contact*. Carl was 46, Ann was 31 and they were described here as 'inseparable'. At one point Carl is quoted saying that before he met Annie, he thought love was hype, designed to sell movie magazines to teenagers.

Their appreciation for each other was tempered by a near-fatal illness to Carl following appendicitis in March 1983,

---

* It drew the largest audience of any programme in US public television history.

which led to massive internal haemorrhaging and 10 hours of surgery. In the same year, Carl made multiple television appearances arguing for a freeze in the number of nuclear warheads in America. Biographers talk about the aforementioned 'Annie effect', and certainly she not only steered Carl towards public campaigning on issues about which they both felt strongly, but she also helped him build bridges with his children from previous marriages.

Five years after *The People* profile, we meet them in the pages of the *NY Times*, this time during a publicity push for just-published *Contact*, Carl's first novel. It describes their multi-level house, with stairs set in the cliffs of Cayuga Heights and a spiral staircase up towards a latticework balcony. Ann is described as tall, chic and 'given to enthusiasm'. Carl's oldest son, Dorion, is also described – he happened to be visiting from Fort Lauderdale.

Today Ann still talks, writes about and champions the themes at the heart of Carl's work: that we are a species poised on the brink of cosmic citizenship. Their relationship has been mythologised and woven into the Voyager story. She shared the 'for keeps' conversation and EEG recording with WNYC's Radiolab podcast in 2007, appeared on numerous panels, discussions and celebrations during Voyager's 40th birthday in 2017. She remains friends with her confidante, producer Lynda Obst, another veteran of Nora Ephron's party back in 1974, and together they're working on a film telling the Annie and Carl story with a working title of *Voyagers*.

Publicly she and Carl were a writing team. Ann inspired the main character in *Contact*, and would co-author many of Carl's most successful works. She co-wrote *Cosmos* and the reboot series that first aired in 2014, and she continues to oversee new editions of Carl's works, preserving his legacy and archive.

'He [Carl] was an unflagging protector,' she says to me in 2017. 'In the early days we'd be in a meeting together and I would gather up the courage to actually speak, and I would say something and someone cut me off right in the middle. And of course Carl was always the alpha male in the room and he would turn and say: "Annie, I believe you were speaking and I really want to know what you have to say." And you know, that other person who had been so foolish to interrupt me, you could see them cringe and sink down into their chair, wanting to disappear and that you know ... He really gave me the confidence to raise my voice ... I kept thinking to myself: "Well, he thinks I'm smart, and he's one smart person, so I really do have something to say." That was one of the infinite number of gifts of being with him.'

In the evening of 20 December 1996 Linda and Ann spoke to each other for the first time in almost 20 years. 'The night that Carl died ... she said something to me and I said something to her that ... immediately meant that we could have a single family united again. And that's how it's been ever since.

'I still feel very guilty about it ... I'm so proud that [Nick] chose to live with his family in the house just at the top of my road, so we could be as close as possible. And I'm really proud that his mother Linda and I are now called matriarchs of this family, inseparable, indivisible. And it will have healed all of the anger and guilt and resentment that inevitably followed.'

Ann wrote later how Carl faced his death with great courage, never seeking any refuge in the 'illusion' of organised religion. In 2012 she wrote about how during their two decades together, they loved each other with constant appreciation of their great good fortune in finding each other in the cosmos, knowing time would be limited.

So let's leave them on their unofficial honeymoon. Let's imagine Christmas in London in 1977. Let's imagine cold grey streets, shops aglow with decorations, snatches of Christmas standards emanating from car radios, coat collars up, hands in pockets, terrible late-1970s British restaurant food. Let's imagine arm-in-arm walks, a hotel room strewn with notes as Carl prepares for the Royal Institute Christmas Lectures. Let's imagine Carl reading passages, Ann making comments. Let's imagine Ann and Carl right at the start of their journey together, a journey that has become part of the Voyager record lore – a story about two very nice human beings united, with 20 years of happiness ahead.

'That was fun. We were staying at Brown's Hotel across the street and we had so much fun ... It was just like a honeymoon. We were beginning to conceive and write our outline for the *Cosmos* television series, the first one ... Throughout our 20 years together there was never any separation between fun and work. We never worked on anything we really didn't believe in wholeheartedly. It was just a feast of ... no borders, no boundaries between our creative life, our family lives, and Carl's scientific life. It was just fantastic.'

★★★

There is another love story bound up in the fabric of the Golden Record, one that hasn't been told so often. Amahl was working at the National Academy of Sciences in Washington DC in 1976 when she met Frank for the first time. She was on a committee. Her committee was organising *another* committee, on which Frank was serving. His committee was tasked with studying employment opportunities in astronomy. She had been corresponding with him for a while, but thanks to an operetta named *Amahl and the Night Visitors*, where the 'Amahl' in question is a young boy, Frank had assumed he was corresponding with a man.

Frank's committee arrived in DC to present their findings to Amahl's committee.[*] It was in a boardroom, underneath a portrait of President Lincoln signing the charter that established the National Academy of Sciences in 1863, that they met for the first time.

Frank recruited Amahl to work with him at Cornell. 'I moved there in 1976,' says Amahl. 'Voyager happened in 1977. We got married in 1978 at the Greek Orthodox Cathedral in Washington DC. I became the stepmom to three wonderful sons, Steve, Rippy and Paul. We built our

---

[*] Amahl was on the staff of the NAS Committee on Science and Public Policy (COSPUP).

dream house and lived in Ithaca until 1984. We were blessed with two beautiful daughters, Nadia and Leila.[*]

After Leila was born, in 1981, Amahl switched careers to become a stay-at-home mom. The family moved to Santa Cruz in 1984, when Frank took a position as Dean of Natural Sciences at UC Santa Cruz.

Frank served as president of the SETI Institute, founded in 1984, for many years. Although he has now retired, he remains Professor Emeritus of Astronomy and Astrophysics at UCSC, and Emeritus Chairman of the SETI Board of Trustees.

Towards the end of my conversation with Ann, I asked her about Frank. She couldn't remember the first time they met, but she said simply: 'I adore him … He was a great friend to Carl. He's the genius behind the cover. That's his. Completely. The scientific hieroglyphs are something that he had been talking about for 15 years before we were able to make the record. He's very modest. And … he doesn't really get credit for his work on the photographs. He's a very brave scientist in that … to pursue these questions of the possibility of extraterrestrial life and intelligence, when that was so looked down upon, and held with such contempt, by the scientific community. He's a guy who can make his own wine, can fix jewellery … he does everything and all with such modesty. He is a marvellous person and he deserves much more credit than he is given for his work on this record.'

Now in retirement, Amahl feels like she's gone full circle back to where she started at Cornell, when she was Assistant to the Director, then as now, supporting, managing and organising his work. She says: 'I think our work on Voyager Golden Record broadened our vision of what Earth is, deepened our view of how beautiful and sacred all life forms on Earth are, emphasised the importance to treasure and protect Earth, and strengthened our understanding of how intricate and essential human interactions are with each other

---

[*] Today Nadia is a science journalist with a PhD in genetics from Cornell. Leila is a prima ballerina with State Street Ballet and a double-major graduate from UC Santa Barbara.

and with our planet. It has made us better people. So, yes, it is nostalgic looking back. It was a profound experience to share with someone who would later become one's life partner. It was a major inspiring event in our lives.

'One of the most striking revelations to me is the fact that, beside Frank and Carl – two of the most noted astronomers of our time – ordinary folks, like me and others on the various teams working on the Golden Record, got a chance to contribute and participate. I don't think this would be true if the record were being made today. Today there would be multiple committees, subcommittees, red-tape, bureaucracies, government or private groups competing to manage the project, travel, expenses, meetings, conference calls, debates, maybe even protests … and most likely team members appointed from a who's-who list. So for me it was the gig of a lifetime to have had a seat at this table.'

She says: 'We are still here. We live in a beautiful home among the fabled redwoods of California. We brought with us our memories from Ithaca and transplanted them here in Aptos where we are getting older and looking back at a life of many accomplishments both professionally and personally … Life is very good.'

Frank remains the popular face for the search for extraterrestrial life. And for every scientist who likes the optimism of the Drake equation, there's the pessimists' charter of the Fermi paradox[*] – the deafening silence, the lack of evidence for extraterrestrial life that has been found to date, despite the seemingly high probability of extraterrestrial life. But, as Frank has repeated in numerous interviews, while the scientific community has been searching for some years now, all that's been done to date is merely scratching the surface. We not only need to search for signals from a million stars, we also have to *keep* looking at them, keep our eyes trained and our ears open for any signs of life.

★★★

---

[*] Named after Italian-American physicist Enrico Fermi (1901–54).

There was one final matter to take care of. The Voyager
Golden Record team had to write a book. Sagan was already a
big-hitting, best-selling, non-fiction publishing phenomenon
by then. So, towards the end of 1977 and early 1978, they
each wrote about their experiences in a series of essays. Frank
wrote 'The Foundations of the Voyager Record', which
traces its origins back to the Pioneer plaques, his own Arecibo
message, some of the technical challenges that faced the
picture team and the design of the record's cover. Jon's essay
focuses on the experiences and rationale of the picture team,
detailing every single image selected (and bemoaning some
that didn't make it). Tim gives an account of the studios at
CBS, before a detailed commentary on every track. Ann
describes the compiling and mixing of the 'Sounds of Earth'
section, right down to individual audio ingredients. There's a
short but fascinating essay from Linda, focusing on her work
on the Cornell greetings, including a chart (the brainchild of
Shirley Arden) showing handwritten transcriptions of each
foreign-language message[*] with a translation alongside. All
this is bookended by two pieces from Carl, charting his
own experiences on the project and looking ahead to what
the Voyager probes might achieve. The book has copious
illustrations, diagrams and photographs (including one
showing an insect flying upside down), plus appendices that
include Robert Brown's letter to Carl, Jon's proto-playlist,
and the final tracklist, as it was known to the team back
in 1977.

   Murmurs of Earth was extremely valuable to me in writing
this book. It represents testimony from the entire Voyager
team written very soon after the events. Like the record itself,
however, it was a hard book to classify: is it a science book or
an art book?

---

[*] This includes a facsimile of six-year-old Nick Sagan's youthful
hand, plus Amahl's message in Arabic. She writes: 'My father wrote,
in his wonderful calligraphic style, the words I recorded, and his
penmanship is preserved in the *Murmurs of Earth* book.'

In any event, Vlado was sitting at his desk back at CBS when Russ Payne walked in brandishing a copy. This was in 1978, months after they had finished the Voyager project.

'He brings me a book. I say: "Russ, what is this? You never gave me a gift before." You know ... "What the hell is this all about?" I'm sayin'. And Russ says to me: "I got you *the book*. This is the book Carl Sagan wrote about our record. And *you're* in it too." And I'm happy, you know? He thanked CBS, he thanked Russ, he thanked me. It was just unbelievable. So I took the book home, you know? And I had the book over there on my shelf, proudly displaying it.' Vlado pauses for a second. 'Unfortunately I went through a divorce, a few years later. And ... I don't know what happened to the book.'

★★★

It was while Voyager 2 was journeying between Saturn and Uranus that I turned from an irritating little third-child snot-bag, into an irritating little boy. I was at a school called New Beacon, in Sevenoaks, Kent. It was an all-boys school, and my best friend was a guy called Michael Ross who's now a pretty amazing guitarist but back then was just another idiot. We both played in the school orchestra. I was part of the two-strong 'third violin' team. When performing the theme tune to *Black Beauty*, us third-stringers handled four notes.

In 1985, there was all sorts of coverage and build-up as Halley's Comet approached. Here in the UK there was a *Horizon* special on the subject. Then there was the actual fly-by, when various space agencies sent an array of probes towards the comet, including the European probe Giotto (named after the Italian Renaissance painter Giotto di Bondone who saw Halley's Comet in 1301, inspiring his depiction of the star of Bethlehem), which went closest of all, photographing a comet's nucleus for the first time before being destroyed in the dust of its tail. I watched the *Sky at Night* special on that too.

By now I had a telescope (it was terrible, but I liked it). I had astronomy books. I had posters of the moon, the solar system and Mars on my bedroom wall. I had a purple ring binder where I drew diagrams, constellations, jotted down definitions, and ruminated over some bizarre, zero-evidence theories of my own. My favourite sci-fi film at the time was *The Forbidden Planet*. Then Voyager 2 reached Uranus.

The closest approach was on 24 January 1986. And a few days later, on a Thursday at 8p.m., Maggie Philbin, Judith Hann and the rest of the *Tomorrow's World* team were all there on my parents' TV, going through the initial results of the encounter. Then there was the 30-minute *Sky At Night* special, 'Voyager to Uranus', on Tuesday 4 February, with Patrick Moore reporting from mission HQ in Pasadena. Then in late May, there was another episode of *Horizon*, picking over the coals of this strange sideways-tipped world with its thick blankets of impenetrable, occasionally rippling cloud, detailing all the things Voyager 2 had told us and the thousands more questions it was posing.

In September 1986, just months after Uranus, and still listening to *Cats* and *Starlight Express* on my Sony Walkman, I was sent to boarding school, aged 10 and still wearing mum-chosen 'civvies' – corduroy trousers and a maroon sweatshirt with 10 white sheep arranged in a pyramid shape across the chest. My mum headed home. Voyager headed on to Neptune. And I headed off to puberty.

I was absolutely obsessed with Voyager's Uranus fly-by. But by the time she reached Neptune (closest approach 25 August 1989) I was a different person. I wore train-track braces on both sets of teeth. I had spots and mood swings. I was hamstrung by anxiety and an all-consuming desire for popularity and cool and acceptance and all that other boring stuff. I wasn't a sporty child, so for my little slice of cool I mined music. There's nothing uncool about astronomy, of course, but by 1989 The Cure had become my primary concern in life. Photos of Robert Smith had replaced my posters of the moon and the solar system. My Cure T-shirt (with lots of spidery 'Lullaby'-era lettering) was the first piece of clothing in which I felt comfortable.

The Neptune closest approach came almost exactly one month after my first concert – The Cure at Wembley Arena,

specifically the final two dates of the Prayer Tour, on 23 and 24 July 1989.*

As Voyager cleared Uranus I sang descant parts in the treble choir, I played the trombone, I wore a blazer and tie to school, held my father's hand around London museums and wore a blue M&S cardigan my mother picked out for smart family events. By Neptune, because of Michael J. Fox and *Slippery When Wet*, I had left my trombone stranded on Grade 3 and taken up guitar instead. I had joined my first band, doing Jesus and Mary Chain covers so drenched in reverb they sounded like they were being played from underneath a pond. I had drunk booze swiped from liquor cabinets and smoked cigarettes stolen from the school matron. I had endured the most painful experience of my life to date – a solo performance of 'American Pie' to students and parents of my boarding house.† I had suffered an unrequited crush on a girl called Amber.

I still remember the Neptune fly-by, don't get me wrong. I still remember staying up for *Sky At Night*. But I remember The Cure's first song in that late-summer evening at Wembley Arena much more vividly. The delicate, chiming bells of 'Plainsong' lasted an age – much longer than the normal LP version. I was wearing a black paisley shirt. I was standing next to my friends Stephen Manotai and James Gossage. The noise of the crowd grew, expectation at breaking point, a wall of ecstatic cheering in near-total darkness, a few pinprick lights on the stage the only indication that band members had taken up their positions. Then a split second of hi-hat preceded an almost physical wall of chunky synth, which erupted from the stage at the same time as a blinding torrent of turquoise and blue-green light – like Voldemort going apeshit – lit us up, row upon row of tiny copycat Robert Smiths, haircuts ablaze. I have goosebumps thinking about it right now.

---

* I went twice. The second was even better than the first because they played 'The Figurehead' and 'Fire in Cairo'.
† It was obvious, from about the second verse, that my housemaster, Mr Sutcliffe, hadn't realised my intention to perform the *entire* song, complete and uncut.

And this all brings me back to why I decided to write this book. Generally, when I used to contribute to *Record Collector* magazine, I loved writing about weird records. Not weird bands so much, but weird records. Strange formats, obscure one-offs, novelty music, forgotten follies. You ever heard of those playable postage stamps that came out of the Kingdom of Bhutan? They were tiny circular stamps, made of a kind of embossed plastic, that could be played on a turntable. Look them up, they're really weird (search 'Bhutan Record Stamps'). Anyway, I wrote about them. You remember Marvin from the original *Hitchhiker's Guide to the Galaxy* series? Well, he released a novelty pop record and I wrote about that too. And, because not everyone I spoke to at *Record Collector* magazine circa 2007 had heard that NASA once sent two records into deep space, right around the time of the 30th anniversary, I wrote about the Golden Records too. The story was this wonderful little nexus, a place where science and astronomy rubbed up against art and music. A place where I could be happy.

During the course of writing this book, I came across a lot of opinion. People love the Voyager record, and people like debating different aspects of it. Some scoff at the whole thing, but most are inspired, or at the very least interested.

There are lots of myths about the record. According to Tim at least, the whole Beatles 'Here Comes The Sun' story has become something of a myth. And there are others. You often hear that 'Satisfaction'* is on there, and I've heard people discuss the project as if it was actually a loudspeaker on the side of Voyager, blaring out music like the stereo of Elon Musk's orbiting roadster.

The Voyager record is certainly a strange story. What I like about it is the way the record seems to have a life of its own. How, despite all our best efforts, it captured something of our imperfections. I like that there's love on there and that, like real love, it's messy. I like that it has some rough edges. I like that the wasp is flying upside down. I like the unscripted Cornell greetings. I like that NASA insisted Jon alter the outline haircut of the man standing beside a pregnant woman

---

* All the fault of the film *Starman*.

so he looked less like a surfer. I like that there's a digitised printout of the names of a load of American politicians on there. I even like that there's a Nazi on there. Indeed, the further away it travels, the more I like it. And I love that these records are hitchhiking aboard these two amazing probes, machines that sprinkled our lives with wonder.

There are mistakes. And you could argue the whole thing, at times and in places, has a last-minute, rushed atmosphere. It has pages that seem almost literally torn out of coffee-table books to represent humanity. It has a picture of three people pouring food and water into their faces to show how we eat and drink. It has weird, mysterious numbers all over some diagrams of human anatomy because Linda's paint flaked off. It has whale song mixed with nonsensical (to an alien audience) greetings. It has an 'h' wearing a backwards cap.

All of these criticisms one could send after the Voyagers, but they won't care. They will carry on regardless, drifting forever in the big nowhere, with their golden discs time-stamped '1977'. The year of punk,* the year of VHS, the year Grateful Dead kicked arse, the year Carl Sagan and Ann Druyan fell in love, the year before Frank and Amahl got married, the year of Apple II, the year of *Star Wars: A New Hope*, the centenary of Thomas Edison's phonograph invention, and the year Peter Carl Goldmark died, developer of the modern LP. And finally the year this disparate team of artists, writers, engineers and scientists put together a monument for humankind – something that will in all likelihood survive longer than Earth itself – and sent it off into space under the noses of some bureaucrats.

Finally, I'd like to end by encouraging all of you to do what I did when researching this book. I had several playlists going. I had the Voyager record as a playlist, which I listened to over and over as I walked my dog along the Sussex Downs. Then, to immerse myself in the mindset of the Voyager team, I decided to make my own. I gave myself 90 minutes to play with and, just as I had for Beth back in the early 1990s, I tried

---

* Well, the year *after* punk really.

to create the *ultimate* mixtape. This was the best of the best. My very, very favourite and bestest music of all time. The music I'd want to preserve, the songs I'd run back into a burning Earth to save if the Sun went nova.

On my first go, I set some ground rules. I limited my pool to music created prior to the launch of these space probes, music that could, in theory at least, have been open to the Voyager team – basically anything released before the summer of 1977. So the Carpenters' version of 'Calling Occupants of Interplanetary Craft' was out as it wasn't released until early September 1977 – just missing the boat – whereas the original version by Klaatu, which first appeared in 1976, was absolutely fine. Then I removed any rules and imagined myself in charge of a record due to leave our planet in a few months' time. The result?

Let's picture an alien being. It has headphones on. It is sitting next to a golden record. Not the NASA Golden Record, but the JSCOTT Golden Record. The alien has a slightly faraway look on its long thin face. It has evidently been sitting there for some time. Suddenly, you can hear a kind of rasping sound coming from the headphones. You notice the stylus is skipping along the run-out groove. A second alien, who's been staring down at a console nearby, looks up at her colleague. The first alien lifts the tone arm, carefully placing it gently back on the armrest. Then he removes the headphones and rubs his pointy ears. He turns to the second alien and says: 'They only had three chords.'

It was hopeless. It was my favourite music, but it was such a narrow view. It was like trying to explain colour by only showing red, or like describing Earth by only showing England's rolling fields. Beautiful, sure, but what about mountains, oceans, lakes, rivers, fjords, deserts, cityscapes, moors, tropical islands, Arctic tundra, savannas, jungles?

Most people reading this will, I hope, be in awe of what the Voyager record team created. They will wonder at Frank's incredible brain. They will be won over by Carl's undoubted magnetism. They will be impressed by Tim's unflappable cool, Ann's infectious wonder, Jon's obsessive dedication, Linda's artistic skill, and the hard work of all the people who left their impressions on the record – Wendy, Val, Amahl, Shirley, Eck, George, all those ethnomusicologists, the CBS engineers, the guys at Colorado Video, the thinkers from SETI and the Order of the Dolphin. But for those of you who think you could do better, I encourage you to give it a try. Here's your brief:

Make a record that represents humanity on a good day.

You can do it in 90 minutes of music.

You can do it in a 12-minute sound essay.

You can do it in 120 images and diagrams of your choice.

You can record greetings in as many languages as you like. And you can put it all on a metal record, inside a metal box.

Don't feel you have to limit yourself to pre-1977 music or photographs. Choose whatever you like, from whenever you like. Mine the world as you see fit.

You have six weeks.

Oh, and before you start, let's just even up the playing field. No computer. No smartphone. No internet. No quick emails. No handy templates. No PDF attachments. No keyword searches. No digital files. No editing software. Go analogue. You're allowed to write letters and make phone calls. You can visit libraries and bookshops. You can trawl catalogues and directories and indexes. You're allowed slides and couriers, darkrooms and film. You can use tapes and records and mixing desks and video cameras.

Good luck with that.

This book isn't about your golden record, or my golden record, it's about *the* Golden Record. A heavy metal album by an awesome band, a supergroup of artists and scientists, supported by a host of ordinary yet exceptional people, who together created this wonderful yet genuinely weird monument. And it is weird. In the best sense of the word, it is weird. It's an odd artefact. A spinning metal plaque with a videoed photograph of a dinner party and the recorded sound of an initiation ceremony. It has the fingerprints of love and enthusiasm all over it, with occasional smudges of darkness. A product of its time. An *objet d'art* partially shaped by bureaucracy, budget, politics and ambition, but filled to its brim with innovation and beauty. It's a metal dream of a half-remembered world. A love letter to us and to them.

Select Bibliography

Murmurs of Earth: The Voyager Interstellar Record, first published by Random House in 1978, was extremely valuable to me and was a source used throughout the writing of this book. It represents contemporary testimony from the entire Voyager team, written within months of launch. It contains three essays by Carl Sagan ('For Future Times and Beings', 'The Voyager Missions to the Outer Solar System' and 'Epilogue'), Frank Drake's 'The Foundations of the Voyager Record', plus essays from Ann Druyan ('The Sounds of Earth'), Timothy Ferris ('Voyager's Music'), Jon Lomberg ('Pictures of Earth') and Linda Salzman Sagan ('A Voyager's Greetings'). It also includes illustrations, a list of all the music, greetings, sound effects and pictures. Appendices include NASA press releases, the UN greetings, a letter from Robert Brown, Jon Lomberg's hour-long playlist, and a list of Voyager mission personnel.

Chapter 1: The Naked Pioneers

Cocconi, G. & Morrison, P. 1959. Searching for Interstellar Communications. *Nature* 184: 844–846

Drake, F. 1978. The Foundations of the Voyager Record. *Murmurs of Earth: The Voyager Interstellar Record*. Random House, New York

Kraemer, R.S. 2000. *Beyond the Moon: A Golden Age of Planetary Exploration 1971–1978*. Smithsonian Institution Press, Washington

Pyne, S.J. 2010. *Voyager: Exploration, Space, and the Third Great Age of Discovery*. Viking Penguin, London

Sagan, C. 1973. A Message to Earth. *The Cosmic Connection*. Anchor Press/Doubleday, New York

—Salzman Sagan, L. & Drake, F. 1972. A Message From Earth. *Science* 175: 881–884

— 1973. A Message From Earth. JPL Technical Memorandum 33-584, Vol. I: 193–203

Wolverton, M. 2004. *The Depths of Space: The Story of the Pioneer Planetary Probes*. Joseph Henry Press, Washington

Chapter 2: Needle Hits Groove

Bell, J. 2015. *The Interstellar Age: The Story of the NASA Men and Women who Flew the Forty-Year Voyager Mission*. Dutton, New York

Cornell Daily Sun archive: cdsun.library.cornell.edu

Davidson, K. 1999. *Carl Sagan: A Life*. John Wiley & Son, New York

Dethloff, H.C. & Schorn, R.A. 2003. *Voyager's Grand Tour: To the Outer Planets and Beyond*. Smithsonian Books, Washington

Drake, F. 1976. Original plan for Voyager Golden Record. Appendices. *Murmurs of Earth: The Voyager Interstellar Record*. Random House, New York

Drake, N. 2014. 40 Years Ago, Earth Beamed Its First Postcard to the Stars. *National Geographic*, Phenomena: www.national-geographic.com/science/phenomena/2014/11/28/40-years-ago-earth-beamed-its-first-postcard-to-the-stars/

The Farthest. 2017. Documentary. Director: Emer Reynolds

Poundstone, W. 1999. *Carl Sagan: A Life in the Cosmos*. Henry Holt, New York

Sagan, C. 1978. For Future Times and Beings. *Murmurs of Earth: The Voyager Interstellar Record*. Random House, New York

— circa 1944–46. The Evolution of Interstellar Space Flight. The Seth Macfarlane Collection of the Carl Sagan and Ann Druyan Archive, Library of Congress

Wawawhack, Rahway High School student newspaper. Vol. VI, No. 5. 1950. The Seth Macfarlane Collection of the Carl Sagan and Ann Druyan Archive, Library of Congress

Chapter 3: Musos v Scientists

Alan Lomax and the Voyager Golden Records, Association for Cultural Equity, www.culturalequity.org/features/Voyager/

Benson, H. Life Among the Stars. *SF Gate*, 6 September 2007

Davidson, K. 1999. *Carl Sagan: A Life*. John Wiley & Son, New York

Drake, F. 1978. The Foundations of the Voyager Record. *Murmurs of Earth: The Voyager Interstellar Record*. Random House, New York

Ferris, T. 1973. Carl Sagan: Life on Other Planets? *Rolling Stone*, 136

— 1977. *The Red Limit: The Search for the Edge of the Universe*. William Morrow, New York

— Biographical details. www.timothyferris.com

Lomberg, J. Unpublished manuscript

Nelson, S. & Polansky, L. 1993. The Music of the Voyager Interstellar Record. *Journal of Applied Communication*. 21: 358–375

Poundstone, W. 1999. *Carl Sagan: A Life in the Cosmos*. Henry Holt, New York

Sagan, C. 1973–76. Examples from the 'Ideas Riding' file. The Seth Macfarlane Collection of the Carl Sagan and Ann Druyan Archive, Library of Congress

— 1977. Letter to Alan Lomax. The Seth Macfarlane Collection of the Carl Sagan and Ann Druyan Archive, Library of Congress

— 1978. For Future Times and Beings. *Murmurs of Earth: The Voyager Interstellar Record*. Random House, New York

— et al. 1978. Appendices. *Murmurs of Earth: The Voyager Interstellar Record*. Random House, New York

Szwed, J. 2010. *The Man Who Recorded the World*. William Heinemann, London

Chapter 4: Uranium Clock

Bart, C. 1976. Profile: Arden Aids Vikings. *Cornell Chronicle* Vol. 08, No. 2

Brown, R. 1978. Letter to Carl Sagan. Appendices. *Murmurs of Earth: The Voyager Interstellar Record*. Random House, New York

Drake, F. 1978. The Foundations of the Voyager Record. *Murmurs of Earth: The Voyager Interstellar Record*. Random House, New York

Druyan, A. 1978. The Sounds of Earth. *Murmurs of Earth: The Voyager Interstellar Record*. Random House, New York

Ferris, T. 1978. Voyager's Music. *Murmurs of Earth: The Voyager Interstellar Record*. Random House, New York

Lomberg, J. 1978. Pictures of Earth. *Murmurs of Earth: The Voyager Interstellar Record*. Random House, New York

— Unpublished manuscript

Sagan, C. 1978. For Future Times and Beings. *Murmurs of Earth: The Voyager Interstellar Record*. Random House, New York

Chapter 5. Now That's What I Call Music

Druyan, A. 1978. The Sounds of Earth. *Murmurs of Earth: The Voyager Interstellar Record*. Random House, New York

— 2007. Article. anndruyan.typepad.com/the_observatory/

Harker, B. 2011. *Louis Armstrong's Hot Five and Hot Seven Recordings*. Oxford University Press, Oxford

Heinlein, R.A. 1976. Letter to Carl Sagan, December 1976. Reproduced in Jon Lomberg's unpublished manuscript

LaFrance, A. 2017. Solving the Mystery of Whose Laughter Is On the Golden Record. *The Atlantic*. 20 June 2017

Lomberg, J. 1978. Proposed music for record. Appendices. *Murmurs of Earth: The Voyager Interstellar Record*. Random House, New York

NASA. The Voyager Greetings: soundcloud.com/nasa/sets/golden-record-greetings-to-the

Nelson, S. & Polansky, L. 1993. The Music of the Voyager Interstellar Record. *Journal of Applied Communication*. 21: 358–375

Ozma Records: ozmarecords.com

Sagan, C. 1995. *The Demon-Haunted World: Science as a Candle in the Dark*. Random House, New York

Smithsonian Folkways Recordings: folkways.si.edu

Chapter 6. The Hydrogen Key

Drake, F. 1978. The Foundations of the Voyager Record. *Murmurs of Earth: The Voyager Interstellar Record.* Random House, New York

Lomberg, J. Unpublished manuscript

Chapter 7. Berry v Beatles

Druyan, A. 1978. The Sounds of Earth. *Murmurs of Earth: The Voyager Interstellar Record.* Random House, New York

Ferris, T. 1978. Voyager's Music. *Murmurs of Earth: The Voyager Interstellar Record.* Random House, New York

Lomberg, J. 1978. Pictures of Earth. *Murmurs of Earth: The Voyager Interstellar Record.* Random House, New York

Nye, W. 2017. Bill Nye talking about classes with Carl Sagan. Big Think: www.youtube.com/user/bigthink

Sagan, C. 1978. For Future Times and Beings. *Murmurs of Earth: The Voyager Interstellar Record.* Random House, New York

— 1986. Birthday letter to Chuck Berry. The Seth Macfarlane Collection of the Carl Sagan and Ann Druyan Archive, Library of Congress

Unknown author. 1977. Voyager Takes Disc Aloft. *Cornell Chronicle* Vol. 9, No. 3

Chapter 8. Flowing Streams and Firecrackers

Davidson, K. 1999. *Carl Sagan: A Life.* John Wiley & Son, New York

Drake, F. 1978. The Foundations of the Voyager Record. *Murmurs of Earth: The Voyager Interstellar Record.* Random House, New York

Druyan, A. 1978. The Sounds of Earth. *Murmurs of Earth: The Voyager Interstellar Record.* Random House, New York

Ferris, T. 1978. Voyager's Music. *Murmurs of Earth: The Voyager Interstellar Record.* Random House, New York

Lomberg, J. 1978. Pictures of Earth. *Murmurs of Earth: The Voyager Interstellar Record.* Random House, New York

Poundstone, W. 1999. *Carl Sagan: A Life in the Cosmos.* Henry Holt, New York

Salzman Sagan, L. 1978. A Voyager's Greetings. *Murmurs of Earth: The Voyager Interstellar Record.* Random House, New York

Sagan, C. 1978. For Future Times and Beings. *Murmurs of Earth: The Voyager Interstellar Record.* Random House, New York

— & Druyan, A. 1997. Epilogue. *Billions & Billions: Thoughts on Life and Death at the Brink of the Millennium.* Random House, New York

Science Friday. 16 September 2016. 'How To Make A Golden Record' hosted by Ira Flatow. www.sciencefriday.com/segments/how-to-make-a-golden-record/

WNYC Studios. 22 October 2007. Radiolab 'Space' episode. www.wnycstudios.org/story/91520-space/

Chapter 9. Mixing and Mastering

Lomberg, J. Unpublished manuscript

Sagan, C. at al. 1978. *Murmurs of Earth: The Voyager Interstellar Record.* Random House, New York

Chapter 10. The Final Cut

Davidson, K. 1999. *Carl Sagan: A Life.* John Wiley & Son, New York

Ferris, T. 1978. Voyager's Music. *Murmurs of Earth: The Voyager Interstellar Record.* Random House, New York

Poundstone, W. 1999. *Carl Sagan: A Life in the Cosmos.* Henry Holt, New York

Sagan, C. at al. 1978. *Murmurs of Earth: The Voyager Interstellar Record.* Random House, New York

Chapter 11. A Last Supper

Lomberg, J. Unpublished manuscript

Sagan, C. 1978. For Future Times and Beings. *Murmurs of Earth: The Voyager Interstellar Record*. Random House, New York

Unknown author. 1977. NASA 'Voyager will Carry "Earth Sounds" Record' press release. JPL website. www.jpl.nasa. gov/news/news.php?feature=6047

Chapter 12. Hello, We Must Be Going

Barry, R. 2017. How to decode the images on the Voyager Golden Record. boingboing.net/2017/09/05/how-to-decode-the-images-on-th.html

Collins, G. 1985. The Sagan: Fiction and Fact Back to Back. *New York Times*, 30 September 1985

Davidson, K. 1999. *Carl Sagan: A Life*. John Wiley & Son, New York

Druyan, A. 2006. Ten Times Around the Sun Without Carl. anndruyan.typepad.com/the_observatory/

Grinspoon, L. 1971. *Marihuana Reconsidered*. Harvard University Press, Cambridge, Massachusetts

McMurran, K. 1980. His Cosmos a Huge Success, Carl Sagan Turns Back to Science and Saturn's Rings. *The People*, 15 December 1980

Nelson, S. & Polansky, L. 1993. The Music of the Voyager Interstellar Record. *Journal of Applied Communication*. 21: 358–375

Ozma Records. 2017. www.kickstarter.com/projects/ozmarecords/voyager-golden-record-40th-anniversary-edition

Poundstone, W. 1999. *Carl Sagan: A Life in the Cosmos*. Henry Holt, New York

Sagan, C. 1977. Briefing notes for President Carter and Vice President Walter Mondale. The Seth Macfarlane Collection of the Carl Sagan and Ann Druyan Archive, Library of Congress

— 1977. The Royal Institution Christmas Lectures, Carl Sagan, 1977. www.rigb.org/christmas-lectures/watch/1977/the-planets

— 1978. For Future Times and Beings. *Murmurs of Earth: The Voyager Interstellar Record*. Random House, New York

— 1993. Pale blue dot: a vision of the human future in space. The Seth Macfarlane Collection of the Carl Sagan and Ann Druyan Archive, Library of Congress

Voyager 2 Fly-by of Neptune Press Conference, 25 August 1989. www.c-span.org/video/?8864-1/voyager-2-fly-neptune

Voyager Missions, Solar System Image Press Conference, 6 June 1990. www.youtube.com/watch?v=ZQCTgCF8Khk

Voyager Spacecraft 40th Anniversary. 5 September 2017. www.c-span.org/video/?433512-2/voyager-spacecraft-40th-anniversary

Others

Alan Lomax Family Collections at the American Folklife Center, www.loc.gov/folklife/lomax/alanlomaxcollection.html

Bell, J. 2015. *The Interstellar Age: The Story of the NASA Men and Women who Flew the Forty-Year Voyager Mission*. Dutton, New York

Davies, P. 2010. *The Eerie Silence: Searching For Ourselves in the Universe*. Allen Lane, London

The Farthest. 2017. Documentary. Director: Emer Reynolds

Ferris, T. 1989. *Coming of Age in the Milky Way*. The Bodley Head, London

JPL Archives Catalogue description of the Voyager Interstellar Record Collection, 1976-1977: pub-lib.jpl.nasa.gov/docushare/dsweb/Get/Document-533/JPL151. For more information on JPL History and Archives go to: www.jpl.nasa.gov/history/

Sagan, C. 2006. *The Varieties of Scientific Experience: A Personal View of the Search for God*. Penguin Press, New York

Appendix A: The Complete Contents of Nasa's Voyager Golden Record[*]

Part I. Greeting from Kurt Waldheim, Secretary General of the United Nations

Part II. Greetings in 55 Languages

Part III. United Nations Greetings/Whale Songs

Mohamed El-Zoeby, Egypt (Arabic)

Chaidir Anwar Sani, Indonesia (Indonesian)

Bernadette Lefort, France (French)

Syed Azmat Hassan, Pakistan (Punjabi)

Peter Jankowitsch, Austria (German)

Robert B. Edmonds, Canada (English)

Wallace R.T. Macaulay, Nigeria (Efik)

James F. Leonard, United States (English)

Juan Carlos Valero, Chile (Spanish)

Eric Duchene, Belgium (Flemish)

Samuel Ramsay Nicol, Sierra Leone (English)

Wallace R.T. Macaulay, Nigeria (English)

Bahram Moghtaderi, Iran (Persian)

Ralph Harry, Australia (Esperanto)

Anders Thunboig, Sweden (Swedish)

Whale songs courtesy of Roger Payne/Ocean Alliance

Part IV. The Sounds of Earth

Music of the Spheres

Volcanoes, Earthquake, Thunder

Mud Pots

Wind, Rain, Surf

Crickets, Frogs

Birds, Hyena, Elephant

Chimpanzee

Wild Dog

Footsteps, Heartbeats, Laughter

Fire, Speech

The First Tools

Tame Dog

Herding Sheep, Blacksmith Shop, Sawing, Tractor, Riveter, Morse Code

---

[*] Tracklist courtesy of Ozma Records.

Ships, Horse and Cart, Train, Truck, Tractor, Bus, Automobile, F-111 Fly-by, Saturn 5 Lift-off
Kiss

Mother and Child

EEG Life Signs

Pulsar

Sound effects and field recordings courtesy of the Elektra Sound Effects Library, except: 'Kepler's Harmony of the Worlds' (Music of the Spheres) courtesy of Laurie Spiegel Publishing (ASCAP); Earthquake courtesy of David Simpson, Lamont Doherty Earth Observatory, Columbia University; Crickets *(Teleogryllus oceanicus)* courtesy of Ronald R. Hoy, Department of Neurobiology and Behaviour, Cornell University; Birds recorded by James Gulledge, Laboratory of Ornithology, Cornell University; !Kung speech courtesy of Richard Lee, Department of Anthropology, University of Toronto; Morse Code provided by William R. Schoppe, Jr.; Train and Saturn V Lift-off recorded by Alan Botto; Mother and Child provided by Margaret Bullowa and Lise Menn, Speech Communication Laboratory of the Research Laboratory of Electronics, Massachusetts Institute of Technology; EEG Life Signs recorded by Julius Korein, MD, New York University School of Medicine; Pulsar courtesy of Frank Drake.

Part V. The Photographic Sequence

President Carter's message (see note at the end of the appendix)
Typed lists of members of the US Senate and House of Representatives responsible for NASA activities 1, 2, 3 and 4

Calibration circle

Solar location map/Andromeda Galaxy composite

Mathematical definitions

Physical unit definitions

Solar system 1

Solar system 2

The Sun composite, Hale observatories

Solar spectrum

Mercury

Mars

Jupiter

Earth

Earth from space, showing Egypt, the Red Sea and the Sinai Peninsula

DNA structure and replication 1

DNA structure and replication 2

DNA structure and replication 3

Cells and cell division, Turtox/Cambosco

Anatomy 1, World Book

Anatomy 2, World Book

Anatomy 3, World Book

Anatomy 4, World Book

Anatomy 5, World Book

Anatomy 6, World Book

Anatomy 7, World Book

Anatomy 8, World Book

Human sex organs, Sinauer Associates, Inc.

Conception silhouette

Conception, Albert Bonniers; Forlag, Stockholm

Fertilized ovum, Albert Bonniers; Forlag, Stockholm

Fetus silhouette

Fetus, Dr Frank Allan

Male and pregnant female silhouette

Birth, Wayne Miller

Nursing mother

Father and daughter (Malaysia), David Harvey

Group of children, Ruby Mera, UNICEF

Family portrait silhouette

Family portrait, Nina Leen, Time, Inc.

Diagram of continental drift

Structure of Earth's core

Heron Island

Seashore, Dick Smith

Snake River and Grand Tetons, Ansel Adams

Sand dunes, George Mobley

Monument Valley, Shostal Associates, Inc.

Forest scene with mushrooms, Bruce Dale

Leaf, Arthur Herrick

Fallen leaves, Jodi Cobb

Snowflake/Sequoia composite, Josef Muench, R. Sisson

Tree/daffodils composite, Gardens

Winterthur, Winterthur Museum

Flying insect with flowers, Borne on the Wind, Stephen Dalton

Diagram of vertebrate evolution

Seashell (Xancidae), Harry N. Abrams, Inc.

Dolphins, Thomas Nebbia

School of fish, David Doubilet

Tree toad, Dave Wickstrom

Crocodile, Peter Beard

Eagle, Donona, Taplinger Publishing Co.

Waterhole, South African Tourist Corp.

Jane Goodall and chimps, Vanne Morris-Goodall

Bushmen hunters silhouette

Bushmen hunters, R. Farbman, Time, Inc.

Man from Guatemala

Dancer from Bali, Donna Grosvenor

Andean girls, Joseph Scherschel

Thailand craftsman, Dean Conger

Elephant, Peter Kunstadter

Old man with beard and glasses, Jonathan Blair

Old man with dog and flowers, Bruce Baumann

Mountain climber, Gaston Rebuffat

Gymnast, Philip Leonian, Sports Illustrated

Olympic sprinters, Picturepoint London

Japanese schoolroom

Children with globe

Cotton harvest, Howell Walker

Grape picker, David Moore

Supermarket

Underwater scene with diver and fish, Jerry Greenberg

Fishing boats

Cooking fish, *Cooking of Spain and Portugal*, Time-Life Books

Chinese dinner party, Time-Life Books

Demonstration of eating, licking and drinking

Great Wall of China, H. Edward Kim

Construction scene, African

Construction scene, Amish, William Albert Allard

Hut

New England house, Robert Sisson

Modern house

House interior with artist and fire, Jim Amos

Taj Mahal, David Carroll

English city – Oxford, *C.S. Lewis, Images of His World*, William B. Eerdmans Publishing Co.

Boston from the Charles River, Ted Spiegel

UN Building by day

UN Building by night

Sydney Opera House, Mike Long

Artisan with drill, Frank Hewlett

Factory interior, Fred Ward

Museum, David Cupp

X-ray of hand

Woman with microscope

Street scene, Pakistan

Street scene, India

Highway, Ithaca

Golden Gate Bridge, Ansel Adams

Train, Gordon Gahan

Airplane taking off

Toronto Airport

Antarctic expedition, National Geographic

Radio telescope (Westerbork, Netherlands), James Blair

Arecibo Observatory

Page from a book

Astronaut in space

Titan Centaur launch

Sunset with birds, David Harvey

String Quartet, Quartetto Italiano, Phillips Recordings

Quartet score/violin composite

Part VI. Short Musical Segue Featuring a Few Seconds of the Cavatina

Part VII. The Music

1. Brandenburg Concerto No. 2 in F Major, BWV 1047: I. Allegro

Composed by Johann Sebastian Bach. Performed by Munich Bach Orchestra/Karl Richter (conductor) featuring Karl-Heinz Schneeberger (violin). Recorded in Munich, Germany, January 1967. Courtesy of Deutsche Grammophon.

2. Ketawang: Puspåwårnå (Kinds of Flowers)

Performed by Pura Paku Alaman Palace Orchestra/K.R.T. Wasitodipuro (director) featuring Niken Larasati and Nji Tasri (vocals). Recorded by Robert E. Brown in Yogyakarta, Java, Indonesia, on 10 January 1971. ℗ 1988 Nonesuch Records.

3. Cengunmé

Performed by Mahi musicians of Benin. Recorded by Charles Duvelle in Savalou, Benin, West Africa, January 1963. Courtesy of Charles Duvelle.

4. Alima Song

Performed by Mbuti of the Ituri Rainforest. Recorded by Colin Turnbull and Francis S. Chapman in the Ituri Rainforest of the Democratic Republic of Congo, circa 1951. 'Alima Song' from the recording *Music of the Ituri Forest*, FW04483, courtesy of Smithsonian Folkways Recordings. ℗ (c) 1957.

5. Barnumbirr (Morning Star) and Moikoi Song

Performed by Tom Djawa (clapsticks), Mudpo (didgeridoo), and Waliparu (vocals). Recorded by Sandra LeBrun Holmes at Milingimbi Mission on Milingimbi Island, off the coast of Arnhem Land, Northern Territory, Australia, 1962. Courtesy of Sandra LeBrun Holmes and Amanda Holmes Tzafrir.

6. El Cascabel

Composed by Lorenzo Barcelata. Performed by Antonio Maciel and Los Aguilillas with Mariachi. México de Pepe Villa/Rafael Carrión (conductor). ℗ 1957 Musart.

7. Johnny B. Goode

Written and performed by Chuck Berry (vocals, guitar) with Lafayette Leak (piano), Willie Dixon (bass), and Fred Below (drums). Recorded at Chess Studios, Chicago, Illinois, on 6 January 1958. Courtesy of Geffen (MCA/Chess).

8. Mariuamangi

Performed by Pranis Pandang and Kumbui (mariuamangi) of the Nyaura clan. Recorded by Robert MacLennan in the village of Kandingei, Middle Sepik, Papua New Guinea, on 23 July 1964.

9. Sokaku-Reibo (Depicting the Cranes in Their Nest)

Arranged by Kinko Kurosawa. Performed by Goro Yamaguchi (shakuhachi). Recorded in New York City, circa 1967. ℗ 1977 Elektra Entertainment.

10. Partita for Violin Solo No. 3 in E Major, BWV 1006: III. Gavotte En Rondeau

Composed by Johann Sebastian Bach. Performed by Arthur Grumiaux (violin). Recorded in Berlin, Germany, November 1960. Courtesy of Decca Music Group Limited.

11. The Magic Flute (Die Zauberflöte), K. 620, ACT II: Hell's Vengeance Boils in my Heart

Composed by Wolfgang Amadeus Mozart. Performed by Bavarian State Opera Orchestra and Chorus/Wolfgang Sawallisch (conductor) featuring Edda Moser (soprano). Recorded in Munich, Germany, August 1972. Warner Classics UK Ltd.

12. Chakrulo

Performed by Georgian State Merited Ensemble of Folk Song and Dance/ Anzor Kavsadze (director) featuring Ilia Zakaidze (first tenor) and Rostom Saginashvili (second tenor). Recorded at Melodiya Studio in Tbilisi, Georgia.

13. Roncadoras and Drums

Performed by musicians from Ancash from recordings collected by José Mariá Arguedas (Casa de la Cultura) in the Ancash, Region of Peru, circa 1964.

14. Melancholy Blues

Written by Marty Bloom and Walter Melrose. Performed by Louis Armstrong and His Hot Seven. Recorded in Chicago, Illinois on 11 May 1927. Courtesy of Columbia Records.

15. Muğam

Performed by Kamil Jalilov (balaban). Recorded by Radio Moscow circa 1950. 'Azerbaijan S.S.R. – Mugam' from the recording *Folk Music of the U.S.S.R.*, FW04535, courtesy of Smithsonian Folkways Recordings. ℗ © 1960.

16. The Rite of Spring (Le Sacre du Printemps), Part II – The Sacrifice: VI. Sacrificial Dance (The Chosen One)

Composed by Igor Stravinsky. Performed by Columbia Symphony Orchestra/ Igor Stravinsky (conductor). Recorded at the Ballroom of the St George Hotel, Brooklyn, New York, on 6 January 1960. Courtesy of Sony Classical.

17. The Well-Tempered Clavier, Book II: Prelude & Fugue No. 1 in C Major, BWV 870

Composed by Johann Sebastian Bach. Performed by Glenn Gould (piano). Recorded at CBS 30th Street Studio in New York City on 8 August 1966. Sony Classical.

18. Symphony No. 5 in C Minor, Opus 67: I. Allegro Con Brio

Composed by Ludwig Van Beethoven. Performed by Philharmonia Orchestra / Otto Klemperer (conductor). Recorded at Kingsway Hall, London, on 6 October 1955. Warner Classics UK Ltd.

19. Izlel E Delyu Haydutin

Performed by Valya Balkanska (vocal), Lazar Kanevski and Stephan Zahmanov (kaba gaidi). Recorded by Martin Koenig and Ethel Raim in Smolyan, Bulgaria, 1968. 1988 Nonesuch Records.

20. Navajo Night Chant, Yeibichai Dance

Performed by Ambrose Roan Horse, Chester Roan and Tom Roan. Recorded by Willard Rhodes in Pine Springs, Arizona, Summer 1942. 'Night Chant, Yeibichai Dance' from the recording *Music of the American Indians of the Southwest*, FW04420, courtesy of Smithsonian Folkways Recordings. ℗ © 1951.

21. The Fairie Round

Composed by Anthony Holborne. Performed by Early Music Consort of London/David Munrow (director). Recorded at Abbey Road Studios, London, September 1973. ℗ 1976 Erato/Warner.

22. Naranaratana Kookokoo (The Cry of the Megapode Bird)

Performed by Maniasinimae and Taumaetarau Chieftain Tribe of Oloha and Palasu'u Village Community in Small Malaita. Courtesy of the Solomon Islands Broadcasting Corporation (SIBC), formerly the Solomon Islands Broadcasting Services (SIBS).

23. Wedding Song

Performed by young girl of Huancavelica. Recorded by John and Penny Cohen in Huancavelica, Peru, 1964. 'Song of Marriage' from the recording entitled *Mountain Music of Peru, Vol. 1*, SFW40020, courtesy of Smithsonian Folkways Recordings. ℗ © 1991.

24. Liu Shui (Flowing Streams)

Performed by Guan Pinghu (guqin). 'Liushui' by Guan Pinghu from the recording *China*, UNES08071, courtesy of Smithsonian Folkways Recordings. ℗ © 1985.

25. Bhairavi: Jaat Kahan Ho

Performed by Kesarbai Kerkar (vocals) with harmonium and able accompaniment. Recorded in Bombay, India, April 1953. Silva Screen Music America.

26. Dark Was the Night, Cold Was the Ground

Written and performed by Blind Willie Johnson (slide guitar, vocals) in Dallas, Texas, on 3 December 1927. Courtesy of Legacy Recordings.

27. String Quartet No. 13 in B-Flat Major, Opus 130: V. Cavatina

Composed by Ludwig Van Beethoven. Performed by Budapest String Quartet. Recorded at the Library of Congress, Washington DC on 7 April 1960. Produced under licence from Bridge Records.

Please note: You will see that I have put President Carter's message and the typed names at the start of my list of photographs on the Voyager Golden Record. In fact, no one seems sure exactly where they ended up. If it was before the main picture sequence, it would mess with the calibration circle. If it was at the end, it would spoil the musical transition. Ozma Records used the original master tapes to create their new edition of the Golden Record. Watch Ron Barry's decode video (boingboing. net/2017/09/05/how-to-decode-the-images-on-th.html) and you can see that this tape's audio includes the complete sequence of photographs, but is without those governmental extras. Apparently there were no other tapes in the Sony Archive with any image audio. So there are numerous possibilities. It could be that the extra pictures were provided on a separate tape, a tape which has since been lost, and these were simply slotted into the Golden Record before or after the main picture sequence. This would slightly spoil either the calibration circle or the musical transition but, short of time, it had to be done. Another likely explanation is that Carter's message and allied names were placed right at the start, before Kurt's spoken words. A presidential message to the universe, albeit in this odd paper-to-slide-to-video-to-audio format, would seem a fitting way to start the Voyager Golden Record. The other possibility is that they were slotted within the main sequence, but this version of the final picture audio edit is simply missing from Sony's archives. The only way to resolve this question might be to play Tim and Vlado's original lacquers that survive at JPL Archives.

The stylus: A cartridge and stylus was mounted inside the cover next to the record. According to John Casani, this wasn't a purpose-built, space-proof NASA-branded needle, but an off-the-shelf, shop-bought model. He estimates they would have cost no more than $20 for the pair. Tim can't remember the brand, but he thinks they would have been more like $80 each, and were definitely ceramic cartridges, rather than magnetic. 'One always avoids putting anything magnetic on a spacecraft without good reason,' he says.

Appendix B: The Cornell Greetings

Here are some, but not all, of the translated quotes that made it onto the Golden Records:

'May all be very well.' (Akkadian)

'Greetings to our friends in the stars. We wish that we will meet you someday.' (Arabic)

'Peace.' (Aramaic)

'To all those who exist in the universe, greetings.' (Armenian)

'Hello! Let there be peace everywhere.' (Bengali)

'Are you well?' (Burmese)

'Hi. How are you? Wish you peace, health and happiness.' (Cantonese)

'Dear friends, we wish you the best.' (Czech)

'Heartfelt greetings to everyone.' (Dutch)

'Hello, everybody.' (French)

'Heartfelt greetings to all.' (German)

'Greetings to you, whoever you are. We come in friendship to those who are friends.' (Greek)

'Peace.' (Hebrew)

'Greetings from the inhabitants of this world.' (Hindi)

'We are sending greetings in the Hungarian language to all peace-loving beings in the universe.' (Hungarian)

'Hello? How are you?' (Japanese)

'How are you?' (Korean)

'Wishing you a peaceful future from the earthlings.' (Nepali)

'We greet you, great ones. We wish you longevity.' (Nguni)

'How are all you people of other planets?' (Nyanja)

'Greetings to the inhabitants of the universe from the third planet, Earth, of the star Sun.' (Oriya)

'Hello to the residents of far skies.' (Persian)

'Peace and happiness to all' (Portuguese)

'Hello to everyone. We are happy here and you be happy there.' (Rajasthani)

'Greetings! I welcome you!' (Russian)

'We greet you, O great ones.' (Sotho)

'Hello and greetings to all.' (Spanish)

'Greetings from a computer programmer in the little university town of Ithaca on the planet Earth.' (Swedish)

'Dear Turkish-speaking friends, may the honours of the morning be upon your heads.' (Turkish)

'We are sending greetings from our world, wishing you happiness, goodness, good health and many years.' (Ukrainian)

'Good health to you now and forever.' (Welsh)

'Best wishes to you all.' (Wu)

Acknowledgements

My first thank you is to Vanessa. Without you, none of this makes any sense. Thanks for putting up with my mood swings and endless talking. Thanks to Genevieve (8) for creating your own Golden Record for the May 2018 Horniman Primary School science exhibition. Thanks to Rupert (13) and Florence (9) too, for asking pointed questions that quickly made me realise I needed to find out a lot more. Thanks to Kate, my middle sister, for unwavering support and advice, and Annabel, my eldest sister, for being the trailblazer I will always look up to as I did when I was 10. Thanks to Mummy and Papa for buying me that telescope and the Cure tickets.

Thanks to everyone at Bloomsbury – in particular Jim Martin for liking the idea and Anna MacDiarmid and Charlotte Atyeo for making it good. David Pescovitz was amazingly kind and supportive. He put me in touch with Tim, Jon, the guys at Colorado Video, and was very helpful answering lots of knotty technical questions I fired his way, sharing information, data, images that he and the Ozma Records team had uncovered. I suspect the Ozma Records contribution to the lore and history of the Voyager record will be remembered for a very long time.

Thanks to Frank, to Ann and Ann's people, who helped find windows in her busy schedule, to Tim for putting up with my asinine follow-up queries, to Amahl for helping co-ordinate my talk with Frank and for answering my questions about her own involvement, and to Jon for kindly sharing his manuscript. Thanks to Julie Cooper at JPL Archives, and to Stephen Dalton, Vlado Meller, George Helou, Jay Pasachoff, Wendy Gradison, Laurie Spiegel, Wyndham Hannaway, Judd Johnson, Mari Noda and Andrij Cehelsky.

Index

21982319055905